GCSE Coursework Series
Series Editors: Jon Nixon and Mike

GCSE Coursework:
Mathematics

A teachers' guide to organisation and assessment

Susan Pirie

**MACMILLAN
EDUCATION**

First published 1988

Published by
MACMILLAN EDUCATION LTD
Houndmills, Basingstoke, Hampshire RG21 2XS
and London
Companies and representatives
throughout the world

Typeset by TecSet Ltd, Wallington, Surrey
Printed in Hong Kong

British Library Cataloguing in Publication Data
Pirie, Susan
Mathematics: a teachers' guide to
organisation and assessment.—(GCSE
coursework).
1. Mathematics—Study and teaching
(Secondary)—England   2. General
Certificate of Secondary Education
I. Title   II. Series
510'.76   QA14.G7
ISBN 0-333-45295-X

Contents

Editors' Preface

Coursework assessment is now a central feature of GCSE. It is a major component in almost all subjects for the six examining groups. Its significance is made clear in the national and subject criteria, and is endorsed by the Secondary Examinations Council in its 'Working Paper 2: Coursework Assessment in GCSE'. Under the old system, coursework sometimes featured in both CSE and O-level: much good and valuable work was achieved, for example, in O-level practical work and CSE mode 3 projects. However, the intentions behind the National Criteria make GCSE coursework quite different from what has gone before. Some element of coursework is now compulsory for all pupils.

CSE and O-level between them were originally designed to examine only 60% of sixteen-year-olds – at that point the least-able 40% were not to sit any examination at all. GCSE is intended to cater for all candidates and, in order to meet the needs of the whole ability range, has introduced the notion of differentiation. In principle this means that all candidates must be presented with tasks which they find manageable and satisfying, and through which they can display positive achievement. Differentiated schemes of assessment are required of all subject areas – and of coursework too. That is, either the tasks set must be closely matched to learners' abilities and competences, or general tasks are set which are then differentiated by outcome – by what pupils actually do or how they perform. In practice, this combination of requirements can make coursework highly interesting and rewarding for the learner. It can also present difficulties of organisation and management for the teacher.

In many ways GCSE coursework seems more prescribed than CSE project work. The onus is now upon all teachers to structure tasks, to set and time them appropriately, and become involved in their assessment and moderation. The purpose of this series is to help teachers tackle coursework within particular subject specialisms. The National Criteria define it as comprising 'all types of activity carried out by candidates during their course of study and assessed for examination purposes'. This means that teachers need to have a clear idea of the aims and objectives of their courses, and the role that coursework tasks play in what they are trying to achieve. The way that coursework is developed within music or drama, for example, will be different from that in science or technology. It is important, too, that the National Criteria say that the 'standards applied in the assessment of coursework must always be those which apply for the final examination, irrespective of when the coursework was actually completed or the assessment made'. That is, a piece of work handed in at the beginning of the fourth year must be judged by the same standards as work completed at the end of the fifth year. It means that teachers must have a clear idea of the quality of work associated with various levels of attainment at age sixteen so that they can gauge coursework at whatever stage in the course it is completed or assessed.

This series addresses the needs of a range of subject areas. Each booklet follows a three-part structure – the first developing ideas and activities in the setting of tasks; the second consisting of a wide range of exemplars of pupils' work, and the third considering issues of assessment and moderation. Parts 1 and 3 serve to inform, raise issues and apply the more general points involved in coursework to each subject specialism. Part 2 consists of original pupil work – sometimes written, sometimes in other forms – and is used as the basis for comments in Part 3. Each author has attempted to address the many variations that still exist between Examination Boards and has attempted to take into account the needs of those teachers engaged in the production of mode 3 syllabuses.

In spite of this common format, the individual books in the series vary considerably. This is partly because the authors bring with them their own distinct perspectives and style; partly because the syllabuses across the various subject areas make very different demands on students and teachers, and partly because of the elusiveness of the term 'coursework'. The National Criteria definition is very broad and its interpretation varies considerably across Boards and subject areas. True, the coursework component is *usually* examined by the teacher and *mainly* undertaken in class, but beyond that there is little consensus as to what might constitute a coursework component.

On the whole, the books are addressed to individual teachers to help in the planning and development of their day-to-day work. However, the assessment of coursework may well depend upon the organisation of courses, and therefore be based upon group decisions or departmental organisation. We hope there is something of value here for groups and subject teams, as well as the lone class-teacher. It is important for all specialists to be aware of the work being carried out in other subject areas. As the boundaries between subjects become increasingly diffuse, we all need to be informed of developments in neighbouring subjects, and of their coursework and assessment needs.

No one can yet be an expert in GCSE coursework – at the moment we are all in the process of learning. As teachers become more practised in task-setting, recognising performance criteria, assessing coursework and undergoing moderation, the whole process will become easier and more familiar. This series is intended to ease the transition towards that stage.

Jon Nixon and Mike Watts

Author's Preface

For some years there has been a gradual shift within mathematical education towards a classroom environment in which pupils are encouraged to create their own mathematics: actively taking part in mathematical thinking rather than passively receiving mathematical thought. The HMSO publication *Mathematics Counts* (1982), better known as 'Cockcroft' after the Chairman of the committee of inquiry, endorsed this change and provoked wide public debate based on its paragraph 243:

> 243 Mathematics teaching at all levels should include opportunities for
> - exposition by the teacher;
> - discussion between teacher and pupils and between pupils themselves;
> - appropriate practical work;
> - consolidation and practice of fundamental skills and routines;
> - problem-solving, including the application of mathematics to everyday situations;
> - investigational work.

With this change of learning has come a shift in emphasis within the curriculum. The paramount importance of the mathematical content of the syllabus has diminished as the acquisition of mathematical processes is seen to have greater value for pupils. The advent of GCSE has set the seal on this change by requiring that, by 1991, all pupils being examined at 16+ must submit coursework for at least 25% of the total mark allowance.

To alter one's teaching style radically can be difficult and threatening. To be forced into such a change accompanied by a completely new assessment system can be paralysing!

One of the aims of this book is to offer support to teachers who are coming to terms with the personal implications of investigative, process-orientated work in their classrooms. Its main purpose, however, is to consider the problem of assessment of coursework as it relates to GCSE. It explores the variety of interpretations of the criteria by the different examination boards and offers activities and suggested reading to help you as a teacher to absorb the new approaches which are being demanded of you. It is essentially a practical book, requiring active participation rather than passive reading. You will want to move backwards and forwards between the various sections, many of which are interdependent and need not be read in strict sequential order. The book's core lies in the central section in which actual pupil work is reproduced, assessed and annotated in order to guide you towards acquiring skills and techniques which allow you to express, with confidence, your professional judgement of a pupil's abilities.

The background and general overview of GCSE mathematics is in the 'red book' (SEC, 1986) which went out to all schools and has not therefore been covered by this text.

Although GCSE is now reality, its form is still evolving and, perhaps for the first time, teachers can be influential in directing future developments. It is vital that you believe in your own ability to assess your pupils' true mathematical understanding and competence. Continuous assessment under normal working conditions by a teacher who knows the pupil *must* be preferable to an externally awarded grade based on a 'one-off' performance under stress.

It is hoped that this book will help you select a syllabus appropriate to the needs of your pupils and will aid your selection of assessment tasks.

Susan Pirie
University of Warwick

Acknowledgements

I should like to thank Rolph Schwarzenberger for his constructive criticism and patient proof reading, Sue Whick who typed through her lunch hours to meet my tight deadlines and Lorna who came to the rescue in a crisis. I am also grateful to all the pupils who provided the materials from which Part 2 was compiled.

The author and publishers wish to thank the following who have kindly given permission for the use of copyright material:

London and East Anglian Group, Midland Examining Group, Northern Examining Association comprised of the Associated Lancashire Schools Examining Board, Joint Matriculation Board, Northern Regional Examinations Board, North West Regional Examinations Board and Yorkshire and Humberside Regional Examinations Board; Southern Examining Group (1988 Examination) and the Welsh Joint Education Committee for extracts from their Mathematics Syllabuses; the Controller of Her Majesty's Stationery Office for extracts from *GCSE: The National Criteria*.

Every effort has been made to trace all the copyright holders but if any have been inadvertently overlooked, the publishers will be pleased to make the necessary arrangement at the first opportunity.

Abbreviations

ATM	Association of Teachers of Mathematics
GAIM	Graded Assessment in Mathematics
LEAG	London and East Anglian Group
MEG	Midland Examining Group
MEI	Mathematics in Education and Industry Schools Project
NEA	Northern Examining Association
NISEC	Northern Ireland Schools Examinations Council
SEC	Secondary Examinations Council
SEG	Southern Examining Group
SMP	Schools Mathematics Project
WJEC	Welsh Joint Examinations Council

PART 1   Task-setting

1   What is mathematical coursework?

The National Criteria (JCNC, 1985) for GCSE mathematics (which are given in full in Appendix 6) set down a list of 15 assessment objectives, 3.1 to 3.15, followed by the statement below:

> . . . Two further assessment objectives can be fully realised only by assessing work carried out by candidates in addition to time-limited written examinations. From 1988 to 1990 all Examining Groups must provide at least one scheme which includes some elements of these two objectives. From 1991 these objectives must be realised fully in all schemes.
>
> (Any scheme of assessment will test the ability of candidates to:)
>
> **3.16** respond orally to questions about mathematics, discuss mathematical ideas and carry out mental calculations;
>
> **3.17** carry out practical and investigational work, and undertake extended pieces of work.

The idea of assessed classwork is not totally new in mathematics – many schools have developed innovative mode 3 assessment schemes – but oral and practical work have, on the whole, been seen as appropriate for the lower ability range, but time-wasting for the more able. The delay until 1991 of the compulsory implementation of these assessment objectives acknowledges that both these aspects of mathematical working will be new to many teachers and pupils. The common image, that mathematics is only about right answers (and neatness!), needs to give way to the idea that creating mathematics involves discussion, exploration, manipulating appropriate materials and, frequently, trial and error or, more formally, hypothesising and testing.

Definition of coursework   The five examining boards*, LEAG, MEG, NEA, SEG and WJEC (see page 8 for the full names of these boards), all intend to realise these objectives through a coursework component in their syllabuses but with a lack of a common definition of 'coursework', each takes a different attitude to what this additional work should be.

NEA is clear that it 'should be an integral part of the (pupil's) study' and 'be undertaken during the final two years of the course'. It should not be 'a sudden and unwelcome burden on candidates and teachers in a few months before the end of the course'.

WJEC stipulates that the work must be done within the year April to April immediately prior to the written examination.

SEG suggests that the chosen work be 'a sample of the results of normal classroom activity rather than additional work carried out merely to satisfy the Coursework requirements'.

---

*The word 'board' will be used as a non-specific label for the new examining groups, associations and councils.

MEG is more prescriptive and insists that each piece of assessed work 'must include an element carried out under controlled conditions . . . e.g. a time or untimed written test . . . or practical test . . .'. This board is undoubtedly the least willing to give teachers autonomy over choice of task and assessment methods. This may be seen as inhibiting and prescriptive or as supportive and helpful depending on the confidence and needs of a particular teacher.

LEAG does not specifically define coursework, but uses the words 'tasks' and 'investigations' apparently interchangeably.

Choice of appropriate coursework

When it comes to describing the specific types of work which may be acceptable for assessment the boards differ quite radically, both in quantity and variety of form.

NEA does not specify any particular tasks, saying, 'The particular form should be chosen for its fitness for purpose in assessing the required aspects of attainment', 'Teachers are free to devise their own assignments' and 'The length of a piece of work may vary from a few minutes' activity to work pursued over several weeks'.

At the other end of the spectrum, MEG states that the candidate must do one assignment from each of the following five categories:

- practical geometry
- everyday applications
- statistics and/or probability
- investigations
- a centre-approved topic

Each task must be the equivalent of two to three weeks' work.

SEG expects three units of work, one of which is an extended piece, while the other two may 'consist of single tasks or collections of shorter tasks' and should 'cover the requirement for practical and investigational work'.

LEAG requires five tasks, one or two from each of the categories:

- pure investigations
- problems
- practical work

WJEC wants only two tasks covering investigations and practical work and they alone differentiate these tasks between levels of examination entry.

The value of oral and aural work

The boards vary also in the third aspect of assessment: the weight given to the components via the mark scheme. In all the following comparisons the percentage mark is that relative to the maximum final examination mark. Assessment objective 3.16 is met in a variety of ways: LEAG allocates 5% for a mental test, but does not specifically consider oral contributions; MEG allows some part of 5% to be based on an oral exchange about the tasks if the teacher desires; NEA allows merely 1.75% to be awarded for oral communication; WJEC gives up to 6.7% for oral assessment and 3.7% for an aural test. SEG attaches more importance to this aspect of assessment than the other boards by offering 8% for oral work together with 10% aural assessment.

Several other bodies, including ATM, GAIM, MEI and SMP (see Author's Preface), are all interested in offering their own particular slant on coursework which could be used as an alternative to the five main schemes. Doubtless, other schemes will emerge. This is discussed further in Chapter 4 of Part 3 of this book.

Activity 1   Achieving the new assessment objectives

Focus on the assessment objectives 3.16 and 3.17. Do any of your pupils already have opportunities to demonstrate an achievement of these objectives? If so, how and when?

Activity 2   Reviewing current schoolwork which might qualify as coursework

Using the headings listed below, consider the work which is at present being done with fourth and fifth-year classes in your school to see whether you could already fulfil some of the coursework requirements of one of the boards as outlined briefly above:
- Extended piece of work
- Investigation
- Practical work
- Practical geometry
- Everyday applications
- Statistics and probability
- Problem-solving

The wide variety of tasks outlined above demands a broad range of assessment methods from the teacher. You need to give thought to what and how you assess.

Activity 3   Exploring the general asssessment demands of different types of coursework

Look at your response to Activity 2 and consider in general terms how you might assess the work. Some of the following questions might help you to focus your thoughts:
- When considering practical work, do you wish to place an emphasis on some finished artefact or on the thinking and planning involved along the way?
- Does modelling form a part of looking at everyday applications of mathematics, or is the focus here to be on that mathematics which is directly applicable?
- When is it appropriate to offer help?
- Do you want to assess an extended project at stages throughout the work, or from an overview perspective at the end?
- What value do you place on accuracy?
- Do you want to assess the mathematics learned through the activity?

The boards all offer guidance notes for use when assessing coursework but these, of course, have differing depth and focus, dependent on the various definitions of coursework. These will be looked at in detail later, but it is important, if the GCSE is not to become the burden referred to by NEA, that you consider your own panics, priorities and preferences in respect to assessment.

Activity 4   Identifying missing elements in current teaching

Consider now the headings for categories of work which you are not at present doing and list some of your inclinations and apprehensions about assessing these types of tasks. Keep these lists. You may wish to refer to them again later.

2   Preparing pupils for assessed tasks

Many teachers are being forced to change their teaching styles to accommodate the need for coursework imposed by GCSE. Pupils, too, must adapt to the new ways of working. Most pupils come to the mathematical classroom with clear images of the likely form the lesson will take and what will be required of them in terms of activities. A Mathematics lesson is assumed to be different from an English lesson or a Physical Education lesson in more ways than just academic content:

- 'Mathematics has no place for personal opinion'
- 'Use of materials other than pencil and paper is only for the primary school'
- 'A calculation is *right* or *wrong*, it cannot be *fairly good*'
- 'Teamwork is not acceptable' . . .

Pupils cannot be expected to throw these ideas away overnight.

Changing ways of working

One of the reasons that the coursework component of GCSE is not compulsory until 1991 is that this will allow teachers to start working in more exploratory ways with their first-year classes so that by the time the pupils reach the examination at 16 they are confident and competent in the new ways of learning. You cannot, however, reasonably expect older pupils to absorb new attitudes 'by osmosis'. You need to make explicit the change in emphasis from content to process. Estimation and prediction (hitherto condemned as 'guessing) are to be encouraged. Checking should not always be done by a teacher but should become a pupil responsibility. Talking to your neighbour is not cheating.

To move pupils towards acceptance of these ideas, persuade them to work in pairs. You may well meet resistance to this unexpected activity. Giving each couple only one piece of paper to work on forces cooperation. When you first introduce investigations or an extended project to a class, even though the pupils enjoy the task, you will meet questions such as, 'When are we going back to proper maths?' or, 'Is this on the syllabus?' Make it overt that you are *not* interested in 'an answer' but in evidence of 'thinking mathematically'.

This book does not attempt to offer help to teachers unfamiliar with taking investigations into the classroom. Pirie (1987) deals with the whole spectrum of changes involved in this shift in teaching style, and this text will assume familiarity with the ideas of investigational approaches and a focus on process. What needs to be made clear is the distinction between allowing pupils the opportunity and freedom to create their own mathematical ideas and explore paths which may or may not be fruitful, and ensuring that they have the skills with which to express their thinking. Moving away from the 'chalk and talk' approach to learning does not mean that exposition and practice are never appropriate.

Oral communication

Three skill areas which become important for the assessment of coursework are *oral exchange, personal recording* and *formal writing up*. Most pupils will be unfamiliar with these skills and will need instruction and practice if they are to do justice to their mathematical achievements. Encourage your pupils to explain to you what they are doing even when engaged in some standard task. 'Talk me through your working', addressed to a pupil who has the *correct* answer, encourages verbalisation of thinking. As you walk round your classroom ask individuals *why?* questions:

- 'Why did you choose those scales on the axes?'
- 'Why do you need to turn thirds and halves to sixths before you can add them?'
- 'Why have you taken a 3 outside the bracket?'

Ask these questions not, in this instance, to probe their understanding, but to encourage them simply to use mathematical words.

A good class activity which emphasises pupils' use of language is for one pupil of a pair to describe a picture, diagram, Lego model or arrangement of counters in such a way that the other pupil can reproduce it without seeing the original. Another task might be to offer the class an erroneous piece of work from a mythical pupil and ask them to explain what is wrong, and why. Do not put words in their mouths; do not say, 'So what you think is. . .'; ask another pupil if he or she has understood or disagrees and promote discussion between the pupils.

Activity 5   Talking mathematics

Select a lesson with a third or fourth-year class in which they will be doing some standard written exercise. Go around consciously asking simple questions which provoke more than just one-word answers. *Listen* to the language they use. Which pupils are having problems putting their thoughts (which you know are correct from their written evidence) into spoken words? When you are doing an example for the class, do you verbalise your thinking, give a running commentary, or just put the working on the board using mathematical symbols? Make 'using spoken language' the main aim for the next few lessons with this class and notice whether their verbal skills improve. You may have a long task ahead of you if the pupils are used to a very traditional way of working, but unless 'encouraging speaking' is one of your overt lesson aims, pupils will neither see the relevance nor acquire the skills for talking about their mathematics.

It is vital that pupils for whom English is not a first language or who have reading difficulties are given verbal mathematical skills which will enable them to express mathematical thinking which would not be exposed in a traditional written paper. The Secondary Examinations Council is quite clear on this point:

> Some candidates . . . suffer out of all proportion . . . because of difficulty in understanding and expressing themselves in written English. Comprehending spoken instructions and questions is . . . a skill worth developing and assessing in its own right; the same is true of oral response. Furthermore, discussion may sometimes give a clearer picture of a child's understanding than a written answer can. (SEC, 1985)

Personal recording

The two further skills, in addition to oral communication, in which pupils need to become accomplished are *personal recording* and *writing up*. These are related but should be distinguished. Neither is normal practice in a traditional classroom. When working on an exploratory, practical or extended project, rough notes, side calculations, *aide-memoires* are all useful, if not essential, yet these conflict with the image of mathematics as some symbolic representation which is brief, neat and, above all, right. How many times have you seen pupils, and in particular girls, writing on the backs of their hands or desk tops before hazarding an answer in their workbooks? Defining recording as a purely personal activity, from which they can later reconstruct their strategies, thoughts and results and from which they can produce a communication to others of their mathematics in the form of a write-up, confers a legitimacy on their rough jottings. Again you may meet resistance and a scepticism that, in spite of all, you *will* 'judge' them by their untidiness and their unconventional recording methods. You have to rebuild a confidence which may have been destroyed many years before.

Writing up

It is the write-up with which you may be primarily concerned when assessing coursework, following this up, perhaps, by a discussion with the pupil. If you intend to judge pupils on planning, strategy, organisation, presentation of results, hypothesising and testing, then you must make sure that they are aware of the importance attached to these processes and work with them to find ways to express them on paper. Two very different write-ups are illustrated for the first task of Part 2, one of which indicates a possible failure on the part of the teacher rather than the candidate. The various aspects on which to judge coursework, as outlined by the different boards, are considered in detail in Chapter 1 of Part 3. Before turning to that section, try this gentle introduction to the new areas for assessment.

Activity 6   Writing mathematics

Plan for your third or fourth-year class a relatively simple investigation which can be substantially completed within one lesson and which is followed by a homework. Focus on the two processes of strategy-building and presentation of results. In the lesson before that in which you offer the investigation, discuss ways of tackling and working through a problem: starting with a simplified situation, drawing a diagram, playing with numbers or materials to gain a deeper understanding. Do the pupils have any other suggestions? Collect together also a list of ways of presenting results: tables, matrices, graphs, drawings, models and so on.

Tell the pupils that their task in the next lesson is to explore a problem and write it up for homework, and say that you will only be looking for evidence of appropriate starting strategies or presentation of results within that write-up. Ensure that every pupil does get started and has some results to record. Resist the temptation to mark the work for grammatical or numerical accuracy and decide what, if anything, you can say about each pupil in the two process areas which are under consideration on this occasion. If you are unable to comment, then the pupil needs further help and practice in communication – since you have arranged that each pupil did in fact, to some degree, perform in each area.

3 Group work and assessment of individuals

The traditional image of mathematics is one of silent, solo working leading to a precise solution or an elegant proof. It is now, however, being recognised that for many people discussion with their peers can often aid understanding; verbalisation helps to sort out ideas as well as to communicate them to others. A natural consequence of the move towards more open ways of working in the mathematics classroom has been an increase in cooperative rather than competitive pupil activity. Tackling an investigation can be a much richer experience if shared by a small group of friends. Ingenious and creative solutions to practical problems can arise from collaboration. Brainstorming is a well-known technique for seeking new approaches and solutions in the business world.

The change from GCE and CSE

Although personal, unaided working is no longer seen as of prime importance to the learning of mathematics, it must be remembered that GCSE is an assessment of the work of an individual pupil. Indeed, public examinations in schools have always played the role of assessing the knowledge, skills and understanding of the individual. GCE and CSE were norm-referenced and by their very definition ensured that the process was also competitive between one pupil and another: only a prescribed proportion of those taking the test were expected to pass. Comparability of results from year to year was held to be all important and fluctuations in the numbers of pupils achieving high marks (as opposed to grades, which were awarded later) were attributed to changes in the facility of the examination paper. It was a system which ranked candidates, but gave little indication of what a pupil could actually do. It could be called a system which distinguished between levels of failure. In such a system, assessment of group work played no part.

Will GCSE be different? Certainly bringing coursework into the arena allows for the possibility of some change from the consideration of only individual pieces of work, although the logistics of how teachers separate individual contributions from a group achievement are at present unclear. What is clear, however, is that the SEC intends that this problem should be addressed:

> The aim in each subject . . . should be one of making what is important measurable, rather than what is measurable important. (SEC, 1985).

This SEC Working Paper on coursework assessment lists the following among the aspects of attainment not easily or adequately tested by a timed, written paper:

(c) Interactive skills (responding appropriately to the consequences of an earlier action); such interaction may be with people. . . .

(d) The ability to find a role and cooperate with others in an activity.

16

Group assignments

At least two of the boards explicitly admit the possibility of group assignments provided the teacher is able to assess the individual candidates' contributions to the work. One suggestion for achieving this is to talk to the pupils individually about the task, in order to find out the roles they played in the activity. An alternative approach is to ask for a personal write-up from each pupil. An example of such a piece of work is included in Part 2. Since both of these activities may be novel for the pupils, they cannot be expected to produce valuable data by which their work may be judged, without some of the practice suggested in Chapter 2. A more innovative approach is currently used in some university departments: where the members of self-selected groups are agreed that the work was truly collaborative a grade is awarded for the work as a whole and each participant is given that grade.

Activity 7   Looking at the effect of measurement on the perceived relative importance of aspects of mathematical learning

Consider how mathematics as taught to 15 and 16-year-olds varies from mathematics as taught to younger children and as experienced in the adult world. To what extent are these differences attributable to the need to prepare for timed, written papers at 16-plus? (from SEC, 1985).

---

4   Differentiated assessment

One radical alternative to norm-referenced assessment is criterion-referencing. This relies upon a set of clearly defined criteria which it is possible to say have or have not been achieved. This is assessment with reference to a body of knowledge in contrast to assessment with reference to a body of people.

Grade-related criteria

The logical end-point of criterion-referencing is a set of grade-related criteria. These would be a series of statements couched in the form, 'The candidate can. . .', which would define each of the grades A to G. Discussion on the production of such grade-related criteria is still very much in its infancy and faces many obstacles. A few general criteria such as 'can select an appropriate starting strategy' would possibly be too broad to be discriminatory, especially since words such as 'appropriate' are open to a whole range of subjective interpretations. On the other hand, use of specific statements such as 'can tabulate results' would lead to an unworkably long list of criteria to be individually satisfied and might lead to meaningless fragmentation of pupils' work. A further stumbling block is that of quality. It would be comparatively easy to say *what* a pupil could do, but far less easy to define *how well* it was done in terms of achievable criteria.

Ways of differentiating

Differentiated assessment is a step on the road to grade-related criteria, where pupils are presented with tasks related to their abilities and are limited as to the grade which they can achieve. The onus for the selection and suitable presentation of coursework lies mainly with the teacher and is all-important when seeking to enable pupils to present themselves in the best possible light. A task which is too easy will not stretch the able pupils and will prevent demonstration of their true abilities. An example of such a piece of work is given in Part 2. Too hard a task will inhibit the less able from displaying those skills which they do possess.

Activity 8   Differentiating the presentation of a common task

Consider the following task which is deliberately phrased in totally general terms:

    'Think of a number. Add *a*. Divide by *b*. Investigate.'

Decide how you would present this to pupils at each of the three entry levels. What structure or freedom would be appropriate? What would constitute success for pupils in each of these groups? How might the help needed differ at each level? What would be a suitable close to the task: a write-up, a discussion, a wall chart, a cut-off at the end of the lesson, an extension to do for homework. . .?

18

The alternative method of differentiating tasks across the ability range is to consider an area of mathematical activity and devise different assignments suitable for each ability group. For example, suppose you wished all the pupils to demonstrate their capabilities through practical work based on scale and measurement. Three possible tasks might be

(a) A rectangular box is 1 m long, 0.4 m wide and 0.5 m deep. Make a scale drawing of the box using 10 cm to represent 1 m. Draw a rough sketch of the net needed to make a model of the box without a lid.

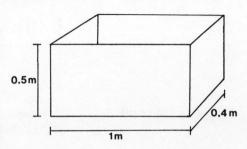

Using the same scale, draw an accurate net for the model. Add flaps where necessary and make the model using glue, *not* adhesive tape.

(b) Measure a football pitch, netball court or other sports area. Using a suitable scale, draw a plan of the area on a sheet of A4 paper. What is the least area of land you would need in order to lay out six similar pitches or courts? How would you arrange them and where would be the best place to site a pavilion?

(c) The Price family wishes to build a bungalow. They want two bedrooms, each at least 9 feet by 11 feet, a dining room, a sitting room, a kitchen and a bathroom. The dining room must be large enough to take their antique table which is $8\frac{1}{2}$ feet by 4 feet and the kitchen must hold a fridge, cooker, dishwasher and deep freeze in addition to the storage cupboards and a sink. Draw a plan of the bungalow using an appropriate scale. Make a two or three-dimensional model and add some model furniture to the rooms.

Activity 9 *Differentiating tasks for a common area of assessment*

Plan different tasks for each of the three levels which will allow you to assess the ability of your pupils to tabulate, spot patterns and generalise. Give careful consideration to the amount and type of help you will wish to give at each stage.

---

5   Collecting and documenting coursework

Involving the pupil

A major concern must be to ensure that coursework is integrated into the curriculum. (SEC, 1985).

Indeed it must, if teachers are not to break under the predicted stress and extra workload caused by GCSE. Some suggestions for ways in which teachers can adapt their normal assessment and marking strategies to include this new aspect are given in Pirie (1987). An alternative response to the SEC statement is to involve pupils in their own assessment. As has already been emphasised, the aspects of attainment to be assessed by coursework must be manifest to the candidates. Pupils should be encouraged to judge for themselves when they have displayed ability in one of these aspects. First, of course, they need to be familiar with the aspects themselves.

Activity 10   Involving pupils in the assessment process

Identify some task which one of your classes has produced and which is in the spirit of GCSE coursework. Ask your pupils to examine their work on this, and possibly other, tasks and to decide which qualities of their mathematical thinking they feel have been exposed by the way they presented their write-up. This can have the dual effect of underlining the value you now place on their active involvement in the assessment of their own work and of sharpening their perceptions of mathematical processes.

Since coursework is to be accumulated over a faily long span of time pupils could be given the responsibility for bringing to your notice mathematical communications in which they feel they have been successful. It is *not* within the spirit of GCSE that assignments should be sprung on pupils as if they were examinations in all but name. It *is* within the spirit that pupils be allowed to submit their work for judgement and that they be assessed on what they can do. It must be said here that the rubric issued by some of the boards is less than clear on this point, although the intention behind the National Criteria and the notion of differentiated papers is that candidates are enabled to demonstrate 'what they know and can do rather than what they do not know and cannot do!' (Forster and Wardle, 1986).

Storage of coursework

One implication of the inevitable accumulation of assessed work is the need to devise a foolproof yet workable method for preserving both pupil work and the teacher judgements passed at the time when the work was done. One of the problems with continuous assessment of adolescents is that they can mature dramatically, both physically and academically, over the two-year period from the beginning of the fourth year to the end of the fifth year. It is not intended that the coursework be produced in the few months prior to the written examinations and therefore success must be judged when the work is produced and not later, with hindsight. This raises an issue in connection with moderation, which is addressed in Chapter 4 of Part 3. It is likely that storage will be an acute problem for school mathematics departments. Practical work and extended projects can imply the accumulation of models, artefacts and large quantities of data. At least until the variety of moderation schemes are seen to work, teachers will be wise to preserve all available evidence to support their judgements. A folder for each pupil, kept by the department, containing each possibly assessable assignment, attached to a brief summary of the teacher's assessment and notes on any further features revealed during discussion or observed while the work was being done, seems the obvious way to preserve written work. Write your thoughts down – you will not remember what was happening in eighteen months' time! In Chapter 3 of Part 3 you are asked to consider this issue in more detail, once you have decided which board you will be using.

6   Feedback to pupils and parents

Summative and formative assessment

Any national examination is, by definition, a summative piece of assessment; that is to say a statement, in this case a public one, of a candidate's ability to perform academically at a particular moment. There is no feedback other than a bald grade, no mechanism for analysis of performance, no opportunity to learn from the event. The intention is to provide the public with a brief statement of where a pupil was at the time of testing. This is an acceptable and indeed proper situation if the examination is in fact taken at a particular moment.

With GCSE, however, assessment takes place across some extended portion of a two-year period. It does not make educational sense for pupils to spend time and effort producing a series of pieces of coursework which disappear into a folder and are never seen by them again! This coursework must be assessed formatively as well as summatively. Pupils should be told how well they perform and where and how they might extend their skills and knowledge. The aim of education is not to produce items for assessment, but rather the aim of assessment is to produce evidence of the benefit a pupil has had from education. Pupils should be involved in learning from their experiences, both in terms of mathematical understanding and of ability to communicate mathematics. Used in this way, the GCSE coursework will not become an extra chore, but will be integrated in the pupils' learning and form one of the tools by which they can judge and improve their own progress.

Just as practical, extended or investigative work calls for a different approach to assessment, so too, assessment calls for different forms of feedback. If the pupil is to build on the understanding gained and appreciate the shortcomings of a piece of work, then marks like '7 out of 10' or 'B+' have no value. Even brief comments such as 'Good' will not help to move the pupil forward. What is needed is a constructive dialogue with the pupil.

Activity 11   Communicating to pupils

Think about the way you are communicating your assessments of coursework to pupils. Do not confuse this with the help given and any discussion you may have while the pupils are working. Focus on your final, overall judgement on a piece of work. You will have graded it in some fashion, but how do you then give the pupils feedback on their performance? Is it given orally or written? Is it brief or detailed? Is it on a regular basis? Are comments addressed to individuals or the whole class? How constructive do the pupils find it? This last question is of real importance if pupils are to advance their learning and performance. If necessary try to devise a more informative and effective way to communicate your assessments on coursework.

Pupils' reactions

You might find it valuable to spend some part of a lesson discussing with the pupils what benefits they derive from your feedback and how they would like to see its form changed. You may well be disappointed to learn that they actually give very little attention to your written comments. Unless they are deliberately involved in the assessment process in some fashion, most pupils, even – or perhaps especially – the more able, see progress in terms of 'learning the content better next time' rather than of pausing to analyse how they approached the problem and what mathematical processes were involved in its solution. Many able pupils are used to 'getting it right' and can be upset by the challenge of an open-ended task. Their analytic thinking powers may not have been tested before and, faced with no obvious 'correct method', they may need much reassurance that 'hazarding a prediction', 'working on a hunch', and most important of all 'being wrong' are necessary adjuncts to real mathematical understanding. A common surprise for teachers when first introducing investigations and practical work is that the less able pupils reveal previously unexplored talents while some of the ostensibly brighter pupils exhibit signals of insecurity and a desire to go back to 'proper maths'. Now as a result of GCSE, should you need such support, you are for the first time able to say, 'This *is* on the syllabus'!

Parents' reactions

If mathematics is looking 'different' to you and your pupils, then pupils' reports of what they 'did in class today' will certainly be confusing and even worrying to many parents. It is essential, if your pupils are not to be torn between two conflicting attitudes, to give parents the maximum information both on the change of emphasis within school mathematics and on the new examination system. A shift towards more 'real-life' mathematics will probably be welcomed, but investigations, particularly those with no 'right answer', will be less easy to comprehend. Parents need to understand that criteria other than just numerical accuracy are valued.

Activity 12   Communicating with parents

Make a list of the areas where the current curriculum differs from the traditional mathematical background, of which the majority of parents will have a hazy and often prejudiced memory. Create a second list of differences of approach and content in GCSE as compared with GCE and CSE. Write a document explaining these in everyday language to be sent out to parents.

Planning a parents' evening

You might consider the idea, used very successfully in several schools, of running a maths evening for parents where they can try out for themselves some of the practical and investigatory tasks on which the pupils are working. At one school this led to parent pressure for the evening to become a regular event!

The kind of activities you plan should be such that firstly they achieve results fairly quickly or provide enough stimulus to incite the parents to stay with the activity, and secondly they use concrete materials in ways that really clarify the underlying mathematics. The parents must be able to see that mathematics as *they* understand it is involved and that their children are not just playing games in their lessons. On the other hand, if parents are to be enticed into being involved, pre-existing mathematics anxiety must not be aroused. Among the areas you will want to illustrate are

(a) practical and extended work,

(b) real-life problems,

(c) oral and language-based tasks,

(d) use of calculator and microcomputer,

(e) investigations using concrete materials or group work.

Consider these in turn.

(a) **Practical and extended work**   Practical work takes time. Extended projects are by definition unsuited to an evening of parent activity. These aspects could be covered, nonetheless, by a display of pupil work, wall charts and artefacts.

(b) **Real-life problems**   You have two constraints to heed here: the mathematics must be seen to be realistically relevant, but it must also be interesting enough to offer a challenge. Carpeting rooms of various sizes is 'real-life' mathematics, but dull. Modelling the problem of laying out a patio with a variety of sizes and shapes of stones could be more interesting, particularly if an assortment of tiles were available to model the problem. A group of parents working on the design of a container for a given quantity of cornflakes, however, and considering the problems of strength, economy of materials, bulk packing and transport, and convenience in the home, is definitely a real-life, stimulating activity, relevant to almost any industry. The challenge addresses parents' common sense as well as their mathematical knowledge of volume, surface area and stability. If scissors and card are provided, the activity becomes a practical one in which prediction and trial are completely acceptable strategies and the benefit of pooling and redefining of ideas which comes with group working is evident.

(c) **Oral and language-based tasks**   Activities centred around Lego are unlikely to appear threatening, yet for one person, having made a model, to describe it to another who must recreate that model without seeing it, demands a precise use of mathematical vocabulary. Similarly describing a diagram so that it can be redrawn, unseen, stretches oral powers of explanation.

(d) **Use of calculator and microcomputer**   Software exploiting the graphical power of the microcomputer and exploring one's mathematical intuitions is appropriate in this context. *Bottles* (JMB, 1985) and *Minimax* from *The Next 17* (ILEA Learning Resources Branch, London) are two such programs. Calculator games must demonstrate the need to think before pressing the buttons. *Four-in-a-line* from *Calculators in the Secondary School* produced by the Open University is a good example of this: most of the arithmetic is actually mental and performed before using the calculator. Definite strategies and use of arithmetic facts are in fact needed to win.

(e) **Investigations using concrete materials or group work**   This will be the least familiar section for most parents. Choose investigations which are easy to start and produce 'results'. *Painted Cube* (Pirie, 1987) is a good example, especially if there are lots of cubes around to work with. The work card approach with an invitation to, say, 'continue the pattern' (Bell *et al.*, 1978) offers a supportive way into generalisation. *Borders* (Pirie, 1987) might encourage discussion of the different ways that people 'see' the border, and lead into algebra as a short-hand notation *with meaning*.

It is important that a member of staff is attached to each activity to offer guidance and support, in addition to promoting discussion, and to draw out for overt consideration the mathematics involved.

Activity 13   Working with parents

Choose with care a selection of activities and plan a parents' evening around them. Your criteria for the suitability of an activity for parents may not be the same as those by which you choose work for your pupils. You also need to decide whether you want a circus of tasks in which the participants dabble and get a feeling for the wealth of possibilities, or an evening where parents become involved, in groups, in only two or three situations, but see the richness of mathematics available in a well chosen task.

PART 2   Annotated examples of pupils' coursework

The following pieces of work have been chosen to illustrate different aspects of coursework and specifically to raise some of the issues connected with oral communication, teacher intervention and writing-up. They have been offered by pupils of widely differing abilities and must not be taken as examples of 'how it should be done'. Except where otherwise stated, the work is that of fourth and fifth-year classes, some of whom have had little previous experience of working and presenting their findings in this way. Each piece of work is preceded by a description of the way in which the task was presented and is accompanied by comments intended to draw your attention to particular features and to suggest questions which you need to consider when looking at the coursework of your own classes. The pieces have not been marked or graded using any of the boards' schemes and it is recommended in Chapter 1 of Part 3 that you perform this activity as a way of familiarising yourself with *your* board's system.

1   A number investigation

The pupils in a top set fifth-year class were each given a piece of paper with the following typed on it:

Think of any 2 digit number.
Multiply it by itself.
Reduce the original number by 1 and increase it by 1 and multiply these two new numbers together.
Investigate.

Compare the two write-ups produced here.

NUMBER INVESTIGATION

Awareness of necessary strategies when pattern seeking.

$12 \rightarrow 144$
$\searrow 143$

I tried one number, but it was not sensible to draw conclusions from this, so I tried another

Appropriate way of recording results.

$10 \rightarrow 100$
$\searrow 99$

At this stage I decided to present my answers in a table.

n	n^2	$(n-1)(n+1)$
12	144	143
10	100	99

Displays sophisticated knowledge of algebraic manipulation.

When labelling the columns for the table, I realised the obvious relationship of

$$n^2 = (n-1)(n+1) + 1$$

Similarly

$$n^2 = (n-2)(n+2) + 4$$
$$n^2 = (n-3)(n+3) + 9$$

I could predict the difference between the two numbers for any change.

$$n^2 = (n-a)(n+a) + a^2$$

A complete generalisation in mathematical terms and justification of the generality.

As the relationship can be represented by algebra, it works not only for two figure numbers, but for any numbers at all.

A clear write-up presenting thinking as well as results. Brief, but revealing high mathematical ability. This pupil demonstrates a rare, powerful economy which is the essence of higher mathematics. One concern, when using a board scheme which prescribes in detail how individual marks should be allocated, is that such a pupil should still be awarded a grade A.

A NUMBER INVESTIGATION

$$27 \times 27 = 729$$
$$26 \times 28 = 728 \qquad \text{one less}$$

Were these the only numbers tried? Can one generalise from 3 examples?

$$134 \times 134 = \cancel{1151} \; 17956$$
$$133 \times 135 = 17955 \qquad \text{one less}$$

$$999 \times 999 = 998001$$

Was a calculator used? — it would have been appropriate.

$$998 \times 1000 = 998000 \qquad \text{one less}$$

It is always one less

Are these conjectures or tried and tested conclusions?

If you change it by 2 it is 4 less

If you change it by 10 it is 100 less

No indication of how the pupil was thinking. No explanations or reasons given.

And so on.

This write-up is an example of a situation where it is not possible to say much, either positively or negatively, about the pupil's ability. This, however, may be the fault of the teacher, if the pupil is unaware of what should be included in a write-up to best display his mathematic talents. (See Chapter 2 of Part 1.)

2 Borders: an investigation

The problem was presented as a written question:

> You are asked to make a border of paving stones around a rectangular pond.
> How many stones will you need?
> What about other shapes?

The pupil chose to work alone.

The investigation involved the number of stones needed to make a border around a rectangular pond.

number of stones needed = $2L_1 + 2L_2 + 4$

where L_1 and L_2 are the lengths of the sides of the pond and the extra four stones are for the corners

The next stage in the investigation said "what about other shapes?" This could mean other shaped stones or other shaped ponds, I presumed it meant ponds.

I considered the following shapes

The initial part of the problem was not, in fact, a problem for this pupil. The answer was obvious to her. Familiarity with mathematical notation is apparent.

L_1

L_2

stone counted twice in perimeter

stone not counted in perimeter

no. of stones = $2L_1 + 2L_2 + 5 - 1$

$= 2L_1 + 2L_2 + 4$

no. of stones = perimeter $+ 4$

30

no. of stones = perimeter + 4

no. of stones = perimeter + 4

Generalisation and
conjectured justification.

Thinking about special
cases which would
test her theory.

The extra number of necessary stones (besides the perimeter) is always
four, as to walk all the way around the pond, one would turn through
360°, which is four 90° corner stones.

I thought about a pond with an island, this would need only
as many stones as the length of the perimeter, so my first idea about the
number of degrees a walker would turn through is incorrect in this case.

A new line of approach.

A new generalisation.

For each corner on the outside of the pond I would need 3 stones, for
each corner on the island I would need one stone, as the outer path really
turns through 270°, and the inner one through 90°

∴ no. of stones = perimeter − 2 × total no. of sides
                    + 3 for each 270° turn of the path
                    + 1 for each 90° turn.

Checking

I made sure that this worked for a complicated border which I already know takes four more stones than the perimeter.

Not completely satisfied with her own proof.

no. of stones = perimeter — 2 x 12 + 8 x 3 + 4 x 1

= perimeter + 4

So it would seem that this does work.

Extending the problem in a new direction

Next I looked at non-rectangular shapes, as below

However, these shapes would require different shaped stones, or stones to be cut for the corners. This could be a whole new problem!

This work is from pupil 1 in the first item. It illustrates the danger of offering GCSE candidates tasks which do not match their ability. Although you might have been struck by the powerful strategy used at the beginning of the problem, based on the realisation that it is sufficient to focus on the corners of the pond to describe the mathematical structure completely, the pupil here has not been given the opportunity to either display to the full her ability to devise appropriate strategies or use higher level mathematical concepts. Differentiated tasks *are* important.

3 Packaging: an everyday application of mathematics, or part of an extended project or practical geometry

The task which follows is a good example of how the same piece of work may be described in different ways and thus fit the coursework descriptions of different boards. It illustrates the fact that, although at first sight the boards are viewing coursework rather differently from one another, in reality pupils on different syllabuses may well be doing similar work.

This class of fifth-year pupils has been working on a cross-curricular project on packaging. This has so far encompassed design in terms of attractiveness and display of information, suitability of packaging materials for different products, and lively discussion of factors such as durability, stability, ease of handling, stacking and bulk packing. The current field of investigation was to consider constraints on the cost of materials.

The task was to produce a popcorn container, with cardboard sides and metal lid and bottom, which was as cost-effective as possible in materials.

Three girls worked together on this problem. Making the popcorn involved weighing, estimating time and following sequential instructions. Although this pupil does not write this up, the existence of the popcorn indicates that that part of the task was successfully completed. Only observation by the teacher, however, would enable a judgement to be made on who was demonstrating which skills. Observation of the group when they returned to the classroom allowed very positive comments to be made on the way they worked together, cooperating and distributing tasks among themselves to make the most of time available. Unfortunately this is not yet a part of any GCSE assessment scheme, although these skills are mentioned in the SEC Working Paper on coursework (see Chapter 3 of Part 1).

Firstly we made some popcorn, then we were given several sheets of A4 sized card and told to find out which shape of box made from the card (the top and bottom were to be made of metal) held the most popcorn. At first we didn't cut the card at all, we just made a circular, a square and a triangular cylinder. By filling each with popcorn we found that the circle circular cylinder held more than the square, and the square cylinder held more than the triangle

We then tried making different sizes of circular cylinders. a taller, thinner one held less, but the shorter and wider the cylinder the more it held (the latter was done by cutting the card into strips and sticking it together, a strip without joins would have been better.) However, from the manufacturers point of view very large, shallow containers would not be practical for two reasons : one that they are awkward to handle and so on, and also use much more metal for the lid & bottom, which is much more expensive than cardboard. Without knowing the relative costs of the card and the metal we could not find out the best combination of height & width from a cost effective point of view.

We were then told to consider the different shapes that could be made without cutting the card further.

We made a octagonal cylinder and a 12 sided cylinder. In ascending order of volume of popcorn they were as follows:-

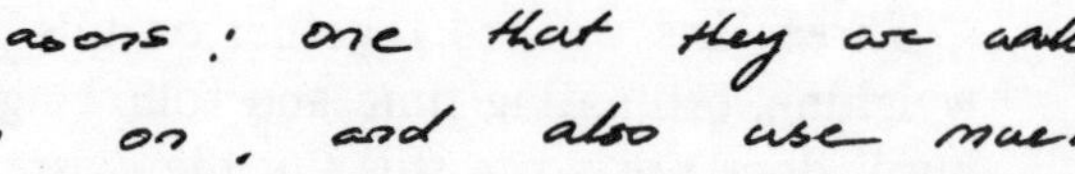

$$\triangle \; < \; \square \; < \; \bigcirc_{(8)} \; < \; \bigcirc_{(12)} \; < \; \bigcirc$$

we therefore decided that the more corners, and therefore the greater the angles involved, the mo

---

Margin annotations:

An appropriate starting strategy.

Appropriate measurement tool (the popcorn)

Suitably modified strategy.

Mathematics and experimentation kept in check by referring the model to reality.

A perceived need for further information, providing a follow-up task for the group.

The teacher re-focused the group as it was not appropriate for them to pursue the information they needed at that moment

Again linking the mathematics and the practical considerations.

greater the volume of popcorn it held, this was because the popcorn wouldn't fit into smaller angle corners and therefore wasted spaces, also a circle is the shape which encompasses the most area for the same perimeter, therefore as the more wider angles there were the closer

A verbal explanation of a mathematical generalisation.

the shape approached a circle the volumes increased.

Lateral thinking.

We also tried a cone, but although this held almost as much as the circular cylinder it was discarded as the metal used would be of an awkward shape and it would also be a difficult shape to pack etc.

Checking practical results with mathematical theory.

We showed that this worked in theory with maths as well as practically :-

s = length of side of square (top surface of $\square$ cylinder)

$\Rightarrow$ the perimeter $= 4s = 2\pi r$

$$r = \frac{2s}{\pi}$$

$\bigcirc$ cylinder volume $= \pi r^2 h$

$$= \pi \left(\frac{2s}{\pi}\right)^2 h$$

$$= \frac{4s^2}{\pi} h$$

$\square$ cylinder volume $= s^2 h$

$\triangle$ cylinder volume $= \left(\frac{1}{2} b h_2\right) h$

$$= \left[\frac{1}{2} \times \frac{4}{3} s \times \sqrt{3}\left(\frac{4}{6}s\right)\right] h$$

$$= \frac{4\sqrt{3} \, s^2}{3} h$$

Mathematical inaccuracy.

The pupil is floundering here due to the conflict between the practical results and the erroneous theoretical solution.

$$\frac{4\sqrt{3} s^2}{3} h \gneq s^2 h > \frac{4s^2}{\pi} h$$

4   Pattern blocks: practical geometry

The pupils were given sets of coloured wooden shapes each with the same length of side. These were:

           triangle – green
          square – orange
        trapezium – red
        hexagon – yellow
     60° diamond – blue
    30° diamond – light wood colour

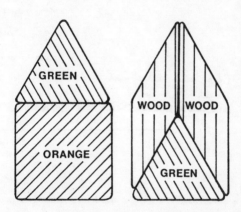

The teacher placed a square and a triangle so that they made a 'house'. She then covered this with two 30° diamonds and another triangle.

The pupils were then asked to investigate the relationships between the shapes. They were allowed to choose whether to work alone or in small groups. Their homework that night was each to write up what they had been doing.

The pupil whose work is reproduced here worked alone. After looking at the homework the teacher (Susan) talked with the pupil, and the transcript of this oral exchange and the subsequent extension of the write-up are also reproduced to demonstrate the value of teacher intervention. It should not be considered here as 'help', which affects the marks awarded, and yet it allows the pupil to clarify her own thoughts and take the problem further, revealing mathematical depth of thinking not apparent so far from the write-up. The teacher did not see the pupil's rough recordings until later, but they are reproduced here since they illustrate a method of recording which was obviously of use to the pupil. In addition they demonstrate a combination of 'home grown' notation and mathematical symbols and tie in with the reference to equations in the transcript.

Write-up

Problem statement and logical deduction which defines the path of her investigation.

The teacher arranged the shapes as below, the two combinations fitting on top of one another, so having the same area. From this one can tell that the square has the same area as two of the narrow diamonds.

 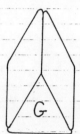

Appropriate method of displaying results

I then tried to find relationships between the areas of all of the shapes. The triangles, trapeziums, hexagons and blue diamonds are related in fairly obvious ways, as shown below.

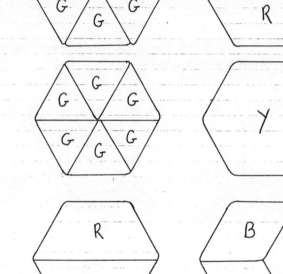

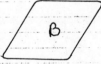 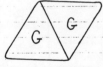

Identifies the heart of the problem.

Next I wanted to know the relationship between the narrow diamonds on the squares, and any of the other shapes.

I thought I had proved something with the combinations below, but was disappointed to see that it only showed what she had shown us; that two long narrow diamonds have the same area as a square.

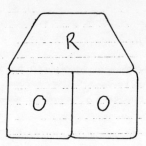
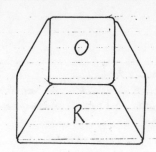

Using a 'trial and error' strategy.

From then onwards I was merely making a reasonable shape and trying to copy it with a different combination. Again, I proved the relationship between the long diamonds and the square, as below.

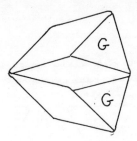
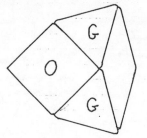

I made one shape which looked good, but didn't prove anything that I didn't already know.

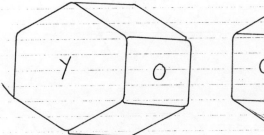
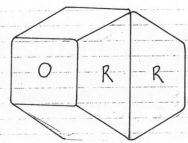

Original strategy proves to be unproductive so a new approach is tried.

The shapes I was making were getting bigger and bigger and not matching until I gave up and used half shapes as below

An explanation of these diagrams (since this is a write-up) would improve communication.

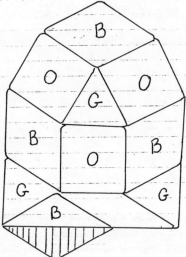
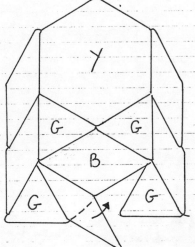

Previous results are used to 'unpack' these two figures.

The Taking from each shape three triangles, one blue diamond and one thin diamond, this left three squares equivalent to four thin diamonds, a triangle and half a blue diamond. Already knowing that two thin diamonds made a square, I took away two squares and four thin diamonds, leaving one square equivalent to two triangles and half a blue diamond. This is the same as a square equalling three triangles.

Next I decided to put together all the information, and realised that I had everything in terms of triangles, so had proved the relationships.

1 trapezium	equivalent to	3	triangles
1 blue diamond	" "	2	"
1 hexagon	" "	6	"
1 square	" "	3	"
2 thin diamonds	" "	3	"

Recapitulation of earlier findings in search of a conclusion.

However, it seems most unlikely that a square is the same size as a trapezium, but I cannot find my mistake.

I thought about using geometry to prove the relationships between the areas of the shapes, but I did not have time.

Perceptive comment and evidence of checking past working.

What she actually said at the end of the lesson was that she thought it was *cheating*, so the teacher said, "write that in", but she didn't! Clearly for this pupil the task was one of working *only* with the materials. This has the real danger that it can result in the pupils appearing to perform less well than their normal capability suggests. As pupils and teachers become more familiar with the new ways of teaching and assessing, this danger will recede, but meanwhile you should be alert to this possibility.

Now turn to the Transcript overleaf.

Transcript

Picking up on the pupil's concern and offering an opportunity to look for an error.

SUSAN: There, I've looked at your homework, I wonder if you want.....You said that you thought that it was unlikely that a square was the same size as a trapezium. I think I agree with you. I wonder if you want to make up that and that diagram and show me what you were doing there.

CATHY: O.K.

S: Which are you making?

C: The left hand one. You see I got a bit stuck there; I couldn't find a way of doing it without 1/2 shapes.

S: O.K.

C: Shall I do the other one as well?

S: Yes, because then you can show me how you matched them off against each other.

C: Well you put yours on top of each other. What I did was put these on top because I was building them out so they matched each other.....There.

S: O.K.You think they're the same?

(referring to the shaded ½ blue diamond and arrow in the diagrams on previous page.)

C: If you take that bit off there and put that on there.

S: So then what did you do?

C: Well I started taking them off each pile to see what I got left with because this one's got some blue in it and so's this one. That one's got some green in it and so's this one. I was taking the same away from each of them.

S: Do you want to do that now?

C: I need to keep the 2 piles separately 'cos its confusing.

S: Then what did you do?

C: I know that two of these brown diamonds make a square, so I can take away 2 blue ones from that one and 4 diamonds from there......And.... Well 1/2 of this isn't supposed to be there.

S: O.K.

C: So then, when I was doing it in my head, I doubled each of these, so I didn't have a 1/2, and put one over there.

S: O.K.

C: I had one of those and two of those..........So that is equal to that.

S: Oh, I see; you've doubled both your remaining heaps.

C: Yes, and then I can take away the 1/2 and put that here. I put that one there because there was a 1/2 there that was supposed to be subtracted. So I called it a whole and put a 1/2 there, like putting it the other side of an equation. Take away one blue thing from there and one blue thing from here. Now I know what those are in terms of triangles: that's 6 triangles, that's 12, that's 14. Each of these is 2. 2,4,6,8, so I can take away 8 from there and 8 from there. So now I've got two squares and a hexagon. I wonder what I had last time.Oh yes, and a hexagon is 6 triangles, so one square is 3 triangles.

S: Right.....I wonder where that...

C: Well you see even that doesn't look right; it looks smaller.

S: It does indeed, doesn't it......

O.K. What did you mean by the bit on the very end? You said "I thought about using geometry to prove the relationships between the shapes." What could you have done? What had you in your head?

C: Well I had already put together some of these brown diamonds, if you put them all nose to nose like that then you can see how many make 360° and then you know how many degrees each one is, and 12 of them go together, so they are 30° each. So I know that and I think 6 of these (triangles) go together, yes, so they are 60° each.

S: So the angles on the green ones are 60° each and the angles on the brown ones are 30° each. Is that what you said?

The teacher is giving No help, or hint of the repeated error.

C: Yes, so.....I thought maybe....I could then work it out making each diamond two triangles and using 1/2 base x height or something, or 1 side x 1 side x sin of the angle.

S: And what do you think that would give you? What shapes would you have to do that for?

C: Well, I know what.... I'm quite sure about the triangles, the hexagons, the blue diamonds and red shapes and I know what they are so I need to do something with the brown ones. If I could link the brown ones to the blue ones then I'd have every thing linked, or something like that. Or the brown to anything out of blue, green or red.

S: How could you find.....

C: They have the same length side as the triangle,but 1/2 the angle I think.....How many angles? You need 6 angles so they're each 60° , so that's 1/2.
They have the same length side so I could work out the area of the triangle in terms of its side and the area of 1/2 the diamond and double it. So if the side is s thats s^2 sin 60° and this one's s^2 sin.....$2s^2$ sin 30°.

S: Fine, I'll leave you and perhaps you'd like to try and write that down.

Pupil is encouraged to work further on the task.

Extension of write-up

The stimulation of talking about the problem has enabled her to spot her two errors without 'help'.

A level of mathematical knowledge and accuracy has now been revealed which completely alters the grade to be awarded for this task.

I have discovered that I was wrong when trying to deal with the blue diamond in subtracting , and infact a square is equivalent to two triangles, which looks more likely, but is not right anyway, as the way I have divided the thin diamond does not fill the hole.

Find the area by geometry:
Let side of triangle be of length s

$$\text{Area} = s^2 \sin 60^{\circ}$$
$$= \frac{\sqrt{3}}{2} s^2$$

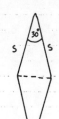

$$\text{Area} = s^2 \sin 30^{\circ} \times 2$$
$$= 2 . \frac{1}{2} s^2$$
$$= s^2$$

A neat conclusion.

There is no whole number relationship between the shapes!

**Pupil's rough
recordings**

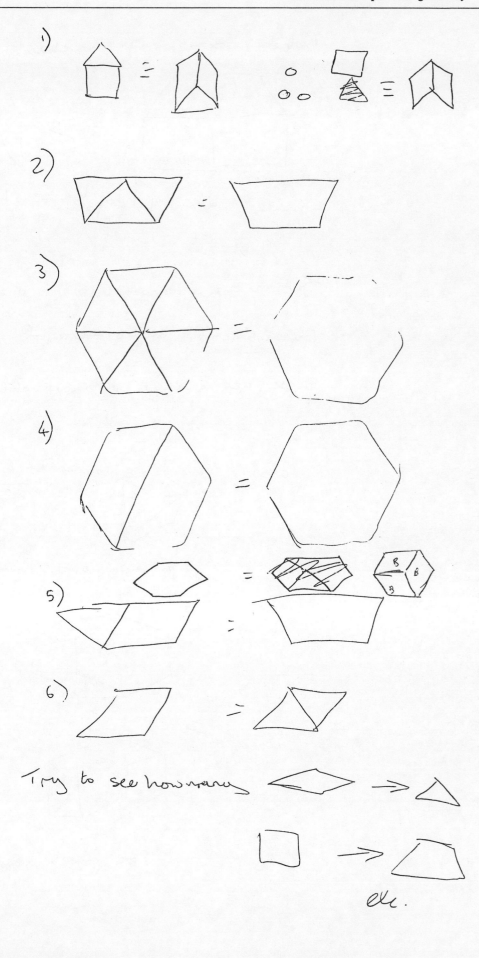

1)

2)

3)

4)

5)

6)

Try to see how many

etc.

Bother ! same as she showed us !

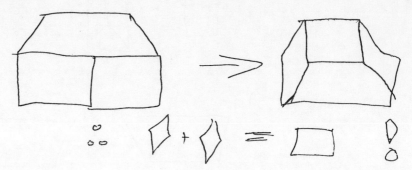

Just making a reasonable shape & trying to copy it.

Trying to match angles - esp of

Shape getting bigger & bigger, still not matched.

also

proved same thing again !

=

Not proved anything -

Have to use $\frac{1}{2}$ shapes!

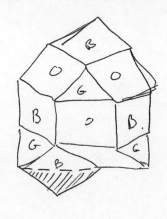

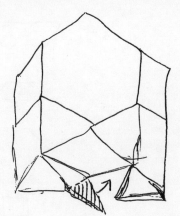

Each minus   3 × △

1 × ▱ B

1 × ◇

3 × ▱B = 1 × ⬡

∴ 3 × ▱ = (4 × ◊) + △

− $\frac{1}{2}$ ◇B

Already know

2 ◊ = 1 ▱

∴

▱O − $\frac{1}{2}$ ◇B = 2 △

⟹   2 ▱ = 4 △ + ▱B

also as   ▱B = 2 △

2 ▱ = 6 △   made a mistake?

= 3 ◇B   ▱ = 3 △

Consolidate info
& realise have everything in terms of $\triangle$'s

⬜ (trapezium) $= 3 \triangle$

▱ (B) $= 2 \triangle$

⬡ (hexagon) $= 6 \triangle$

⬜ (square) $= 3 \triangle$ ⎫
                                    ⎬ But slightly
◊ + ◊ $= 3 \triangle$ ⎭   doubtful about
                                    these

↓

Square tiles smaller than

⬜

5   Painted cube: an investigation

This is the work of a fourth-year pupil who was not in the habit of using concrete materials. The class was presented orally with the problem of deciding how many unit cubes, cut from a larger, painted cube, would have 1 painted face, 2 painted faces and so on. Each table had a large heap of multi-link cubes on it. The pupils sat four to a table, but the majority worked on their own, chatting while they worked. The teacher strongly encouraged the use of the cubes, and the writing up was done during the lesson. From observation it was clear that this pupil enjoyed doing the task, but, unlike the pupils in task 4, was not inhibited by the materials from bringing his other mathematical knowledge to bear on the problem.

The write-up raised some questions in the teacher's mind which were answered to her satisfaction when she looked at the pupil's recordings. Although recording is usually for the pupil's benefit, it can be, as here, legitimate to draw positive conclusions from these recordings, with the pupil's permission.

What is evident from this work is that the pupil needs help in writing down his thinking, not merely his answers. The teacher was surprised to discover that this pupil appeared, from the recording, to have actually built and dismantled each of the cubes without attempting to predict from a pattern, and she talked to him about this:

T Goodness, did you really make all those and take them to pieces one by one?

P Yes.

T Couldn't you have had a guess at what the next one might be?

P Sure. After a bit I knew what they were going to be but I liked building them and then seeing the tally pattern grow.

An examiner who did not know this pupil might not have initiated this snippet of conversation and might have had to judge the pupil to be using an inappropriate strategy with no attempt at prediction or conjecture.

Write-up

I made some cubes of varying sizes and took them apart to see how many 1x1 cubes would have 0,1,2,3,4,5 or 6 sides red if the whole cube had been painted red when it was whole.

These are the results I obtained:

Appropriate method of displaying results

size cube	no of faces red						
	0	1	2	3	4	5	6
1x1x1	0	0	0	0	0	0	1
2x2x2	0	0	0	8	0	0	0
3x3x3	1	6	12	8	0	0	0
4x4x4	8	24	24	8	0	0	0
5x5x5	27	54	36	8	0	0	0
6x6x6	64	96	48	8	0	0	0

A Generalisation with a justification.

As cubes have eight corners the number of 1x1x1 cubes within the larger cube that have 3 faces red will always be 8.

For an $n \times n \times n$ cube these will be the results.

$$0 : (n-2)^3$$
$$1 : 6(n-2)^2$$
$$2 : 12(n-2)$$
$$3 : 8$$
$$4 : 0$$
$$5 : 0$$
$$6 : 0$$

Sudden algebraic leap. Where did this come from? Was it his own work?

provided that $n \neq 1$
$n \neq 2$

Sophisticated use of notation.

Pupil's personal recordings

cube → faces	0	1	2	3	4	5	6
1×1×1	⊘		⊘	⊘	⊘	⊘	1
2×2×2	⊘	⊘	ЖЖ IIII 8		⊘	⊘	⊘
3×3×3	1	ЖЖ ЖЖ 6 11	ЖЖ ЖЖ ЖЖ 12	ЖЖ IIII 8	⊘	⊘	⊘
4×4×4	ЖЖ III 8	ЖЖ ЖЖЖЖ IIII 24	ЖЖ ЖЖ ЖЖ IIII 24	ЖЖ III	⊘	⊘	⊘
5×5×5	ЖЖ ЖЖ ЖЖ II 16	ЖЖ ЖЖ ЖЖ ЖЖ ЖЖ ЖЖ IIII 36	ЖЖ III 36	⊘	⊘	
6×6×6	27 64 ... 64	54 ... 96	48 ... 48	ЖЖ III 8	⊘	⊘	

$(n-2)^3$ $6(n-2)^2$ $12(n-2)$ 86 $6(n-2)^2$
                                                6×25

16
14
6 4
32
$6 \rightarrow 18 \rightarrow 30 \rightarrow 42$
$0 \rightarrow 6 \rightarrow 24 \rightarrow 54 \rightarrow 96$

$\begin{array}{ccccc} & 6\times4 & 6\times9 & 6\times16 & \\ 6 \rightarrow & 24 \rightarrow & 54 \rightarrow & 96 & \\ & 18 & 30 & 42 & \end{array}$

$n \geq 3$
$6(3-2)^2$
6×2
$=12$

$6(4-2)^2$
$6(2)^2$
$6 \quad 4$
24

$6\sqrt{54}$

$6\overline{)96}$ $\;16$

$n\times n\times n \qquad n \neq 1$
$\qquad\qquad\qquad n \neq 2$
$0 = (n-2)^3$

$6 =$

6 Worms: an extended project

This project took a third-year pupil two weeks to complete, including the write-up. For the purpose of assessment, however, treat it as a piece of GCSE coursework and decide at which level you feel this candidate should be entered.

WORMS

Clear presentation
of the problem.

There is a worm which moves in a pattern. It would slither along for, say 2 cm and then turn through 90°. It would then slither for maybe another 3 cm and turn through another 90°. Then it would slither for 12 cm, prehaps, and turn 90° and repeat it's pattern again :- 2cm, turn, 3cm, turn, 12cm, turn, 2cm, turn, 3cm turn, and so on. Every turn would be clockwise if it's first ever turn was clockwise, or every turn would be anticlockwise if it's first ever turn was anticlockwise. This is where we get the idea of the Mathematical worm problem.

Treats and discards
the trivial cases.

Worms usually have 3 or more figures. 1 and 2 figure worms are feasible, but are uninteresting.

1 figure worms.

1 →   This is the pattern of a 1 figure worm.

Generalises accurately.

1 →   This is the pattern of a one figure worm repeated.
If it was repeated again and once again it would form an equilateral quadrilateral or a square

As we are trying to find different shapes and patterns made by worms, I consider the worm with one figure - in one word - BORING.

2 figure worms.

2,3 →   This is the pattern of a two figure worm.

2,3   This is the pattern of a two figure worm repeated. As you can see it forms a symmetrical quadrilateral or a rectangle. All two figure worms, whatever the numbers will form quadrilaterals, when repeated.

Examples of 2 figure worms
forming quadrilaterals, when repeated.

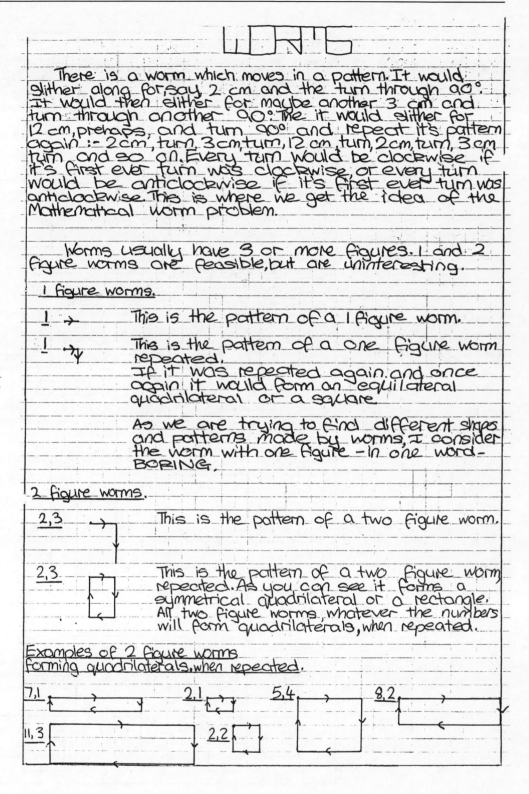

7,1   2,1   5,4   8,2

11,3   2,2

Predicts and checks
Trivial case.

3 figure worms.

3 figure worms form more interesting and exciting
patterns than worms with 1 or 2 figures. The only
uninteresting 3 figure worm is the one with all three
of its figures being the same. If this worm is repeated
it will trace the same steps as before and eventually
on the fourth time of repition it will end up back at
the original starting point.

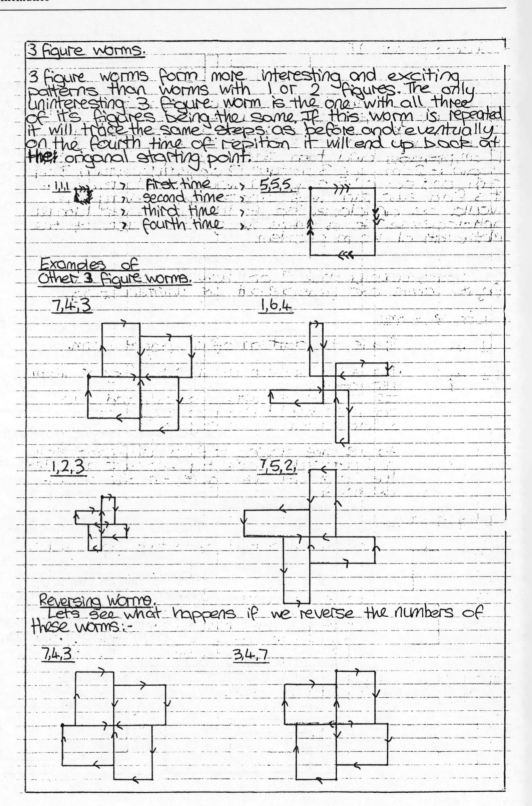

1,1,1 →   > First time   > 5,5,5
          > second time >
          > third time   >
          > fourth time >

Examples of
Other 3 figure worms.

7,4,3   1,6,4

1,2,3   7,5,2

Speciatises in order
to experiment and
get a 'feel' for the
problem.

Reversing worms.
Lets see what happens if we reverse the numbers of
these worms:-

7,4,3   3,4,7

Tries a new line
of investigation.

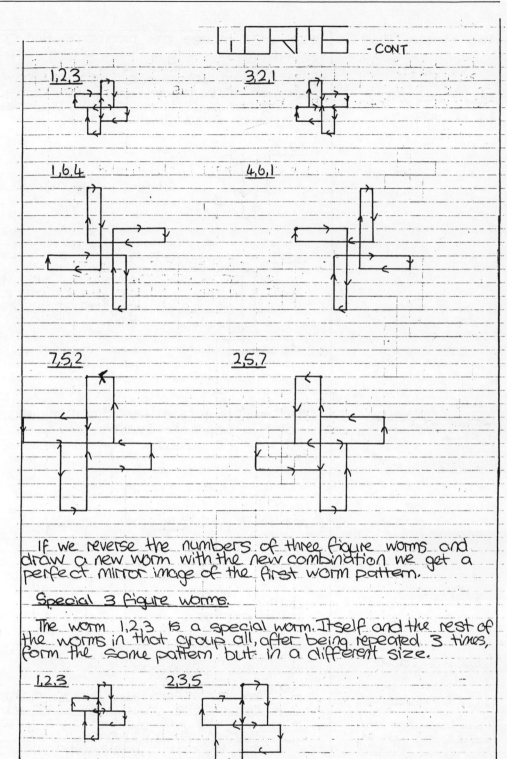

WORMS - CONT

1,2,3   3,2,1

1,6,4   4,6,1

7,5,2   2,5,7

Spots the pattern and generalises.

If we reverse the numbers of three figure worms and draw a new worm with the new combination we get a perfect mirror image of the first worm pattern.

Special 3 figure worms.

The worm 1,2,3 is a special worm. Itself and the rest of the worms in that group all, after being repeated 3 times, form the same pattern but in a different size.

New line of approach, not very clearly explained.

1,2,3   2,3,5

Poses some questions.

Differentiating the drawn patterns.

Looking for number patterns.

Makes a conjecture

based on examples.

Poses a question and predicts the answers, with reasons for the predictions.

But what makes the patterns the same?
And what similarity is there between the numbers?

With 3 figure worms, each pattern, draw when the worm is repeated three times, has four 'arms'. These arms may or may not meet in the middle. In this group of worms the arms all meet in the middle.

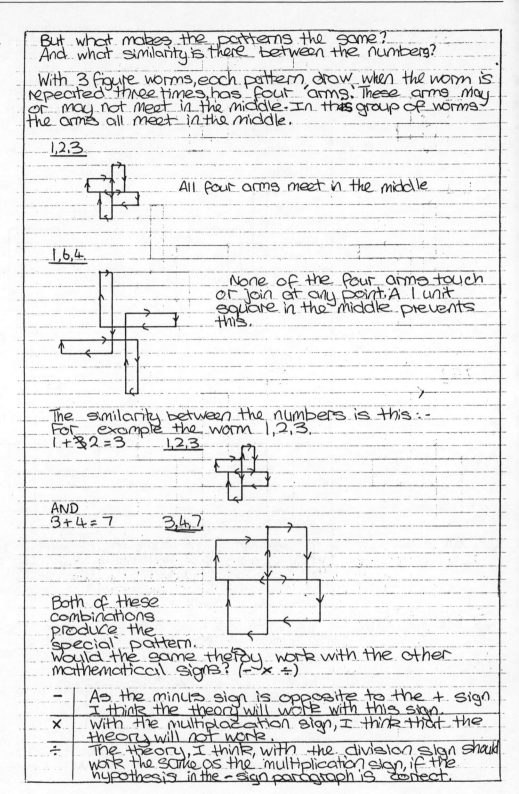

1,2,3

All four arms meet in the middle

1,6,4.

None of the four arms touch or join at any point. A 1 unit square in the middle prevents this.

The similarity between the numbers is this:-
For example the worm 1,2,3.
1 + 2 = 3   1,2,3

AND
3 + 4 = 7   3,4,7

Both of these combinations produce the 'special' pattern.
Would the same theory work with the other mathematical signs? (- × ÷)

−	As the minus sign is opposite to the + sign I think the theory will work with this sign.
×	With the multiplication sign, I think that the theory will not work.
÷	The theory, I think, with the division sign should work the same as the multiplication sign, if the hypothesis in the - sign paragraph is correct.

WORMS - cont.

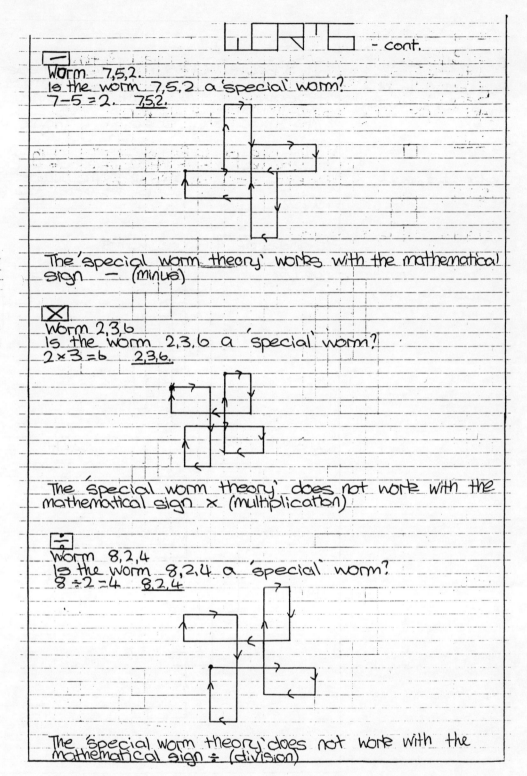

☐ (minus box)
Worm 7,5,2.
Is the worm 7,5,2 a 'special' worm?
7-5=2.   7,5,2.

Is one example enough to generalise from?

The 'special worm theory' works with the mathematical sign — (minus)

☒
Worm 2,3,6
Is the worm 2,3,6 a 'special' worm?
2×3=6   2,3,6.

The 'special worm theory' does not work with the mathematical sign × (multiplication)

÷
Worm 8,2,4
Is the worm 8,2,4 a 'special' worm?
8÷2=4   8,2,4

The 'special worm theory' does not work with the mathematical sign ÷ (division)

Extends the problem.
New question based on
a previous observation.

Looks at special
cases.

Specialises.

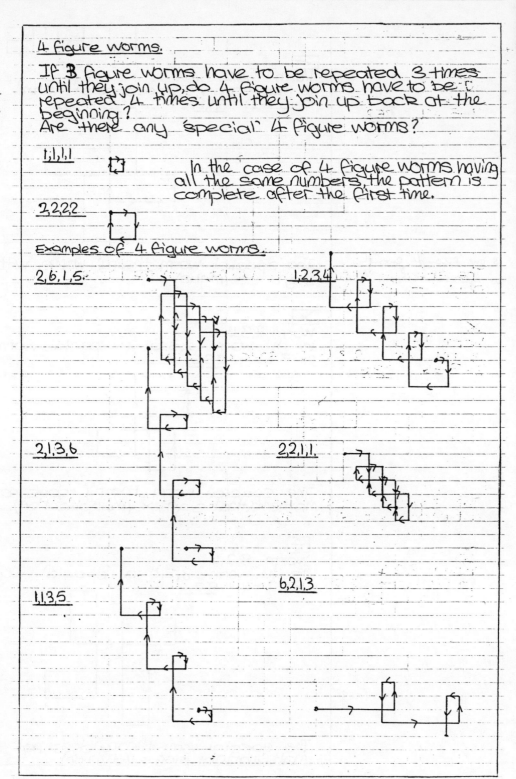

4 figure worms.

If 3 figure worms have to be repeated 3 times
until they join up, do 4 figure worms have to be
repeated 4 times until they join up back at the
beginning?
Are there any 'special' 4 figure worms?

1,1,1,1

In the case of 4 figure worms having
all the same numbers, the pattern is
complete after the first time.

2,2,2,2

Examples of 4 figure worms.

2,6,1,5.   1,2,3,4

2,1,3,6   2,2,1,1

1,1,3,5   6,2,1,3

WORMS — CONT

Special 4 figure worms.

Although 4 figure worms never join up to the point at which they started, do they still have special form similar patterns with similar number combinations?

2,2,1,1.   4,4,2,2.

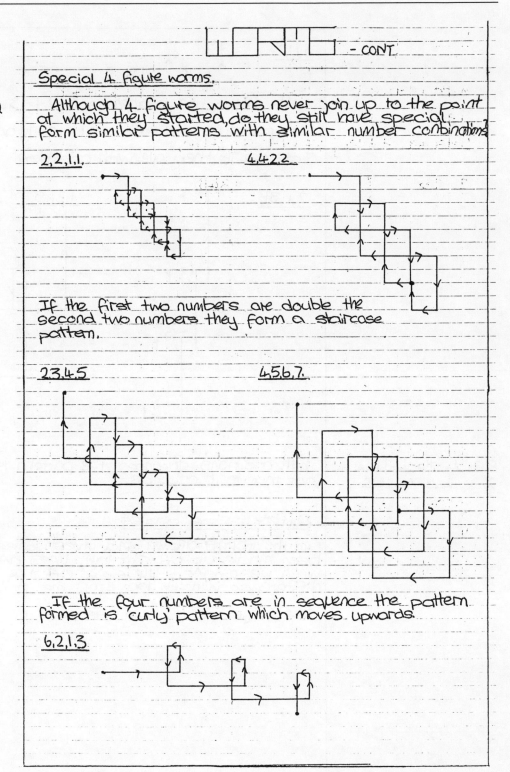

If the first two numbers are double the second two numbers they form a staircase pattern.

2,3,4,5   4,5,6,7.

If the four numbers are in sequence the pattern formed is 'curly' pattern which moves upwards.

6,2,1,3

<u>4,1,1,2</u>

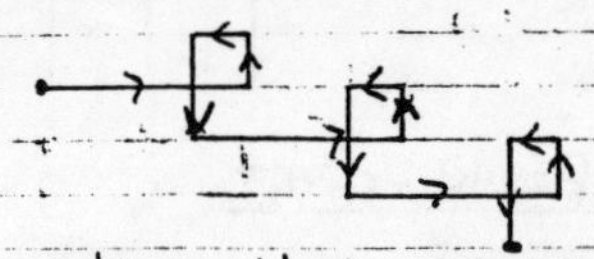

With subtraction sequences 6-2-1 = 3 and 4-1-1 = 2
                              6 2 1 3   4 1 1 2
the patterns are both similar, curly and downward

Lets see if they are with addition.

<u>2,1,3,6</u>   <u>1,3,4,8,</u>

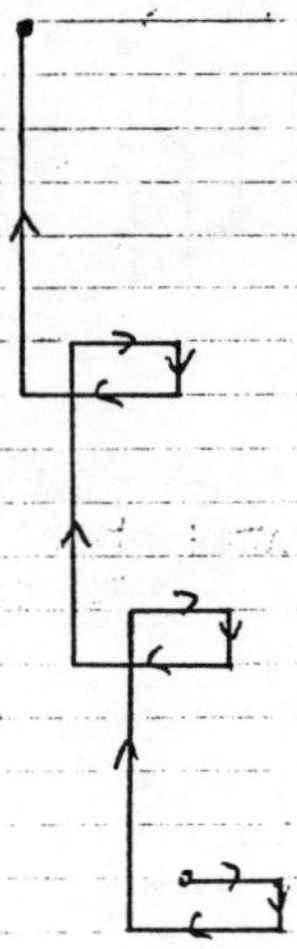

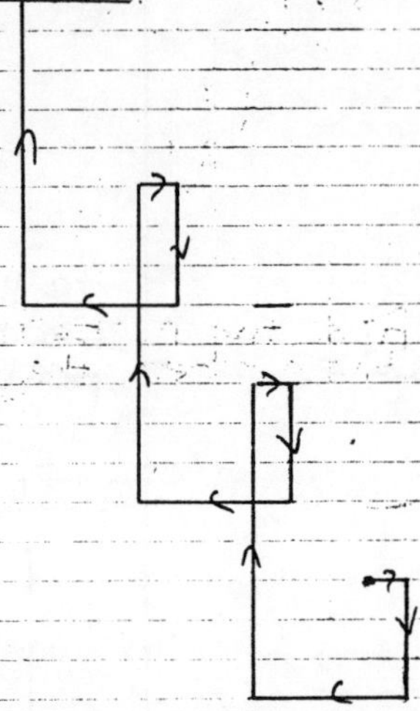

With additions sequences the pattern is similar in the
fact that it ascends.

<u>Different size angles.</u>

All the worms drawn and investigated so far have
been made up of straight lines and angles of 90°.
When we change the size of the angles what
difference will it have on the patterns formed?

The angles we will being using are 60° and 120°.

Extends the problem
in a new direction.

Explores the new
problem.

60°| 1 figure worms   2 figure worms

__1__

__2__

__2,5.__

__4,2__

3 figure worms.

__2,7,5__

__1,1,&2.__   __4,6,3__

__1,2,3__

60° | 4 figure worms.

4,1,5,2.

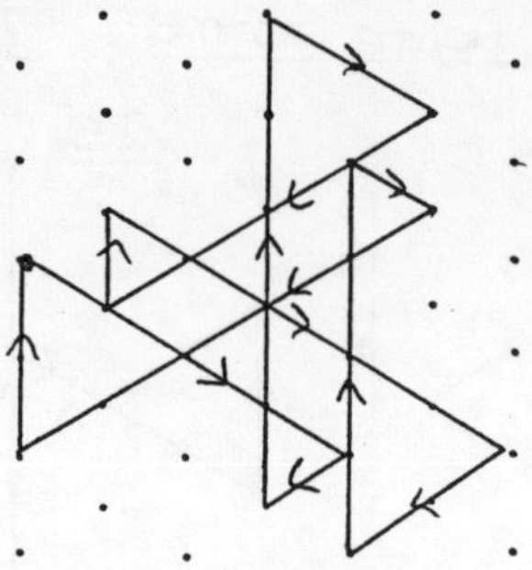

3,4,2,1

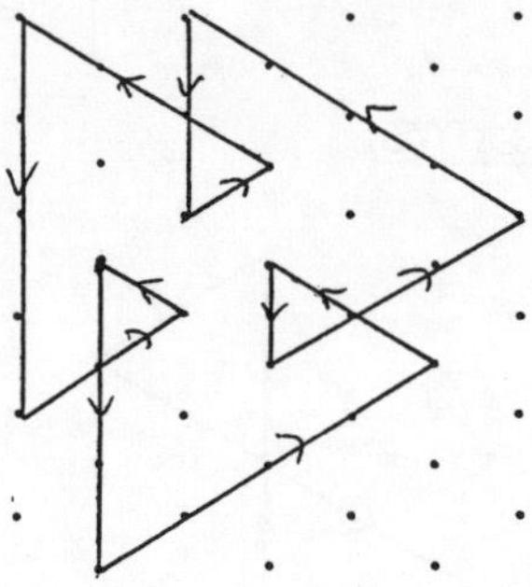

Summary of the results

With angles of 60° :-

		MOVES TO GET BACK TO STARTING POINT
1 figure worms	Form Equilateral Triangles.	3
2 figure worms	Form a shape made up of three equilateral triangles.	3
3 figure worms	Form patterns but never return to the starting point. as with 90° angles	Never gets back
4 figure worms	Form shapes and also return to the starting point, the opposite to with 90° angles	3

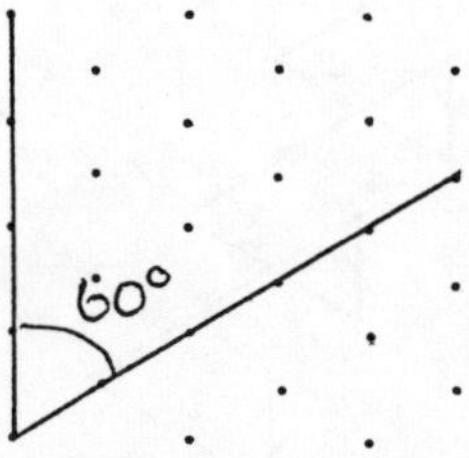

60°

The 60° angle

WORMS – CONT

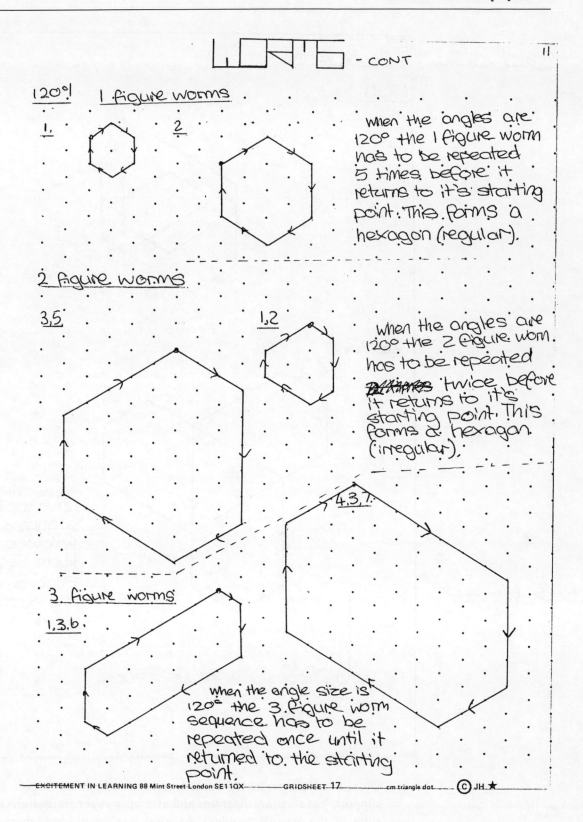

120°!   1 figure worms

1.
2.

When the angles are 120° the 1 figure worm has to be repeated 5 times before it returns to it's starting point. This forms a hexagon (regular).

2 figure worms

3,5
1,2

When the angles are 120° the 2 figure worm has to be repeated ~~times~~ twice before it returns to it's starting point. This forms a hexagon (irregular).

4,3,7

3. figure worms

1,3,6

When the angle size is 120° the 3. figure worm sequence has to be repeated once until it returned to the starting point.

Further extension.

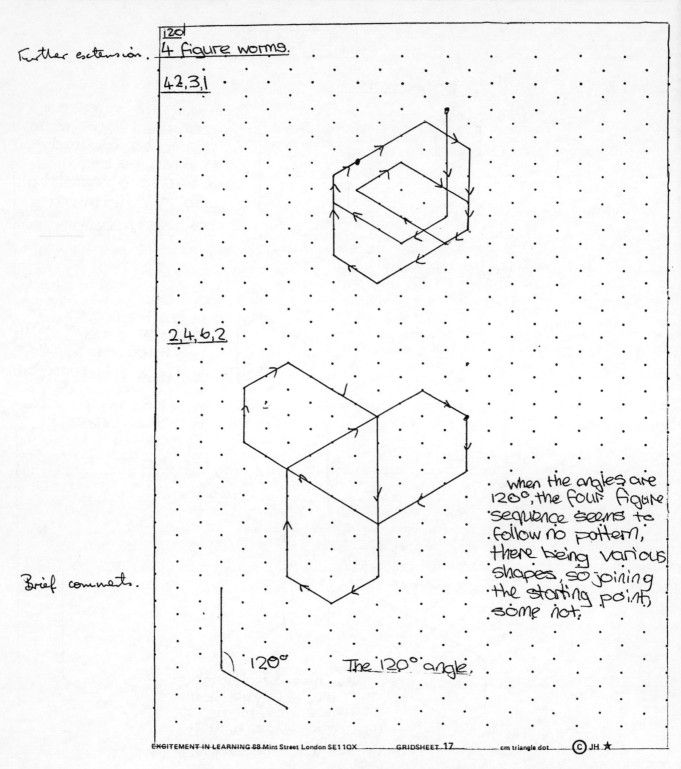

120
4 figure worms.

4,2,3,1

2,4,6,2

when the angles are
120°, the four figure
sequence seems to
follow no pattern,
there being various
shapes, so joining
the starting points
some not.

Brief comments.

120°   The 120° angle.

Overall a well presented record of her work and findings. Little theoretical
support, but accurate diagrams and attempts at generalisations for several extensions of the original problem. At what level would you enter this candidate so
that she can demonstrate achievement?

7 Boxes: practical geometry or investigation

The class was shown how to make a box by cutting a square out of each corner from a 20 x 20 grid and then taping the edges together. They were asked to consider and test out which box would hold the most. No other teacher help was given. Group work was neither suggested nor discouraged. This pupil was seen to be working on her own, although the curious impersonal style of write-up might lead one to think otherwise. One of the dangers of using practical methods with pupils for whom this is not a normal, everyday experience is that they may feel constrained to work only within the limitations of the materials. This was obvious in task 4, but is possibly also true in this case, where the pupil does not follow up her own extension.

<u>BOXES</u>

Boxes were made by taking a 20×20 piece of grid paper
and cutting a square out of each corner and then folding the
remaining paper to form a box:-

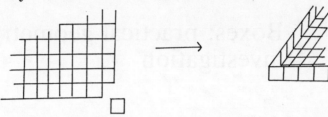

Several boxes were made to find which would note hold the most.
Before making the boxes a prediction was made as to which shape of
box would hold most, a shallow box, a tall and narrow box or a cube-
shaped box. The prediction was a shallow box but not one with
only 1 unit in height.

The boxes were labeled according to the square that had been cut out
ie. box 1 had had a 1×1 square cut out of each corner, box 2 had had
a 2×2 square cut out of each corner and so on.

The volumes of the boxes were recorded and the results were as
below:

Box	Dimensions	Volume (units³)	
1	18×18×1	324	The pattern of the dimensions
2	16×16×2	512	became obvious and the volume
3	14×14×3	588	could have been found without
4	12×12×4	576	making the boxes but the
5	10×10×5	500	corners were enlarged from the
6	8×8×6	384	one before as a check.
7	6×6×7	252	
8	4×4×8	128	
9	2×2×9	36	
10	0×0×10	0	

The box that holds the most, out of the boxes tested, is the 14×14×3
box.

If it was not necessary to cut along the lines a box that could
hold more could be found, the box with a square of 3.5×3.5 cut
out of each corner, this gives a volume of 541.5 units³.

The range of boxes, cutting along the lines, are from 0 units³ (0×
0×10) to 588 units³ (14×14×3), the former holds the least, the latter holds
the most, it holds the most because it has the biggest volume. It
holds the maximum volume because it has a large base area and is
fairly high. The result was similar to the prediction, a shallow box.

This problem might have application in the world when people
are planning how to pack things and in what sized container.

PART 3   Assessment

1   What does assessment of coursework involve?

Interpreting assessment objectives

The variety of interpretations by the examination boards of the assessment objectives and the general demands made on teachers by the need to assess coursework were indicated briefly in Chapter 1 of Part 1. The specific practicalities of implementing these requirements will now be considered. Although there are other options for GCSE coursework being proposed by a variety of bodies, including MEI and SMP, we concentrate first on the five major boards.

The National Criteria for GCSE mathematics include aims and assessment objectives which are given in full in Appendix 6. Each board reproduces these lists with the exception of SEG which rewrites the assessment objectives under the headings of 'Knowledge', 'Skills', 'Applications', and 'Problem-solving' (Appendix 7). This board also has a totally different approach to the whole process of assessment and could be considered the only board genuinely to adopt the positive approach of discerning what pupils *can* do. Each grade from G to A is defined by a set of criteria whereas the other boards merely rely on the grade description for F and C laid down in the National Criteria. The SEG Teachers' Guide states categorically that the 'Group is committed to the ideal of teachers taking responsibility for the shaping of the new system' and that it is therefore 'essential to bear in mind that in essence *all* GCSE literature is still in "draft form" '. This is a challenge which should not go unheeded and should be taken to apply to the literature of every board. The philosophy of SEG is that teachers are professionals who know best the abilities of their pupils and the assessment scheme is built on teachers knowing the appropriate grade for a piece of work, a system which has long been in operation in universities, where tutors make with confidence statements such as 'this is a good 2(i) but not quite a first class'. In the SEG scheme a numerical mark is then awarded for each unit from a table, dependent on the level 1, 2 or 3 at which the candidate has been entered. The other boards all ask teachers to provide a final mark formed from an accumulation of marks awarded for specific aspects of the work and teachers are not expected to allocate a grade.

The attitudes of the boards to the whole idea of entrusting teachers with the task of publicly assessing their pupils' abilities is revealed by subtle distinctions between what is prescriptive and what merely guidance. It must be stressed that this section is not intended to be judgemental: as has already been said, some teachers will feel security lies in a system with a clear marking scheme, while others will prefer the freedom to exercise their informed judgement. The table overleaf gives information on each of the boards' schemes under various headings.

Activity 14   Using different assessment schemes

Study the board comparison table and notice where the different emphases lie. Even if you already know which board you will be using, take one of the pieces of work in Part 2 and award it a 'gut-feeling' grade. Then assess it using two or

Comparison of boards'
assessment schemes

		LEAG	MEG
% Marks	Course:	25	25
	Exam:	70	75
	Other:	Mental 5	

	LEAG	MEG
Papers (All have three levels)	Sit 2 papers from a chain of 4 Level X: Grades E – G Level Y: Grades C – F Level Z: Grades A – D Mental: 3 different tests	3 different pairs of papers Foundation: Grades E – G Intermediate: Grades C – F Higher: Grades A – D
Coursework assignments	5 Investigations. One or two from each of: • pure investigation • problems • practical work	5 Assignments, each 2–3 weeks work One from each of: • practical geometry • everyday applications • statistics/probability • investigations • centre approved topic
Board prescription and guidance on task choice	7 coursework tasks/investigations will be set by the board. Candidates may submit investigations of their own choice provided they fall into the required 3 categories	Topics are suggested for each of the task headings, but 'candidates are encouraged to select suitable topics outside this list'
Allocation of marks	Each task marked out of: • 7 strategy • 7 implementation and reasoning • 6 interpretation and communication Total: 20 × 5 = 100 scaled to 25	For each task there are 4 marks for each of: • overall design/strategy • maths content • accuracy • presentation and clarity of argument • a controlled element (brief test or oral exchange) Total: 20 × 5 = 100 scaled to 25
Board prescription and guidance on marking	The criteria for awarding 2, 5 and 7 marks for Strategy and Implementation and 1, 4 and 6 for Interpretation are given (See Appendix 1)	The criteria for awarding 4, 2 and 0 marks for each section of each task are given (See Appendix 2)
Moderation	A random sample by board's Moderators	A random sample by board's Assessors

(for similar information about NISEC see Appendix 8, page 94.)

NEA	SEG	WJEC
25	40	22.2
75	50	74.1
	Aural 10	Aural 3.7
Sit 2 papers from a chain of 4 Level P: Grades E – G Level Q: Grades C – F* Level R: Grades A – D* *Lower grades awarded in exceptional circumstances	Sit 2 papers from a chain of 4 Level 1: Grades E – G Level 2: Grades C – F Level 3: Grades A – D Aural: 2 from a chain of 4	Sit 2 papers from a chain of 4 Level 1: Grades E – G Level 2: Grades C – F Level 3: Grades A – D Aural: 1 test for Levels 2 and 3, separate test for Level 1
Centres are responsible for deciding nature of coursework 'The length of a piece of work may vary from a few minutes activity to work pursued over several weeks'	3 Units of work One must be an 'extended piece' Other two are 'single tasks' or a series of short assignments	Level 1: An investigation of a particular theme using various mathematical techniques and three exercises demonstrating application of certain practical skills. Levels 2 and 3: A practical investigation and a problem solving investigation
A list of possible activities is offered only as a guide. 'Teachers are free to devise their own assignments'	'The provision of suitable activities is left to the teacher.' Together the units should involve all of the following selection: data handling, problem solving, interpretations and generalisations of results, communication of results	A choice of 2 topics for each task at Levels 2 and 3 will be set. 2 topics and 4 exercises will be set for Level 1. Centres may choose their own topics provided they are of similar standard to those of the board

Practical work: 30
Investigation: 40
Assimilation: 30
 	
Total: 100 scaled to 25
	Extended piece: 12 × 4
	    (weighting)
	2 other units each: 12 × 2
	    (weighting)
	Oral aspect: 12 × 2
	    (weighting)
	
	Total: 120 scaled to 40
		The marks at Levels 2 and 3 are divided between:
		(i) The finished assignment.
		• understanding 3
		• strategy 5
		• content and development 6
		• communication 7
		(ii) Continuous assessment including oral assessment.
		• understanding 3
		• method 3
		• conclusion 3
		Total: 30 × 2 scaled to 22.2

Assessment guide outlines the main points a teacher should consider and the mark ranges which can be used. These include:
• planning   0–12
• skill in using equipment   0– 6
• understanding of equipment 0–12
• clarity and conciseness of
  expression   0–10
• recall of knowledge   0– 3
• understanding of concepts
  and subject matter involved 0–10
(See Appendix 3)
	4 pages of questions to ask either about the work or the pupil are offered as guidance. The teachers' aim is to award a grade first and numerical mark afterwards. Detailed cumulative grade descriptions are given
	
	(See Appendix 4)
		A description of the meaning of each of the headings is followed by the criteria for awarding each individual mark
		
		(See Appendix 5)

Statistically against written paper marks, but poor coursework cannot affect grades adversely (1988–1991)	Consortium consensus or by inspection by the board	Consortium moderation or by inspection by the board

three different schemes, the details of which are in Appendices 1–5. Are the resulting marks the same? Which scheme is closest to your own feelings on the worth of the work? Which scheme is easiest to use? Which scheme do you prefer?

Questions to ask of your board's scheme

The purpose of this activity is to bring to the fore the differences in the various boards' approaches and in their interpretations of the National Criteria and thus to raise some questions which you need to answer about the scheme you will be using. You should be clear whether you are assessing a final product or the ongoing work of a pupil. WJEC, for example, is quite specific that the assessment will include 'continuous assessment including oral assessment during the four-week period' allotted to each task as well as 'assessment of the completed assignment'.

When dealing with the oral component you will need to differentiate between assessment, through an oral exchange, of work done or understanding gained, and assessment of pupils' abilities to talk about their mathematics. These are definitely not the same: in the first case the onus is on the teacher to elicit the pupil's ideas but in the second the onus is on the child to display communication skills. SEG makes it clear that *both* aspects are to be assessed by saying that the oral component shall 'assess the candidate's ability to respond orally to questions about mathematics and to discuss mathematical ideas'. The SEC Working Paper (SEC, 1985) refers to testing the skills of interaction 'with people, information sources, tools or concrete materials' and the place of group work has been discussed in Chapter 3 of Part 1. Some of the boards require an aural test, but this alone does not fulfil assessment objective 3.16. It is, however, one way of testing an ability to 'carry out mental calculations'.

Activity 15   Getting to know your board's requirements

Read carefully the documentation from your board and answer the following questions:

1 When is the assessment performed: before and/or after completion of the work?
2 What, precisely, is the oral component? When and how is it assessed?
3 Is there an aural element? Do you ever do this kind of mental testing with your pupils?
4 How will you incorporate assessment into situations where pupils are working in groups?
5 What does the word 'investigation' appear to mean to your board? Is this compatible with your definition?
6 If you are using one of the more prescriptive schemes, will you be able to recognise when a pupil achieves each mark-awarding criterion? If the allocation of marks is up to you, are you happy that you can work within the guidelines offered?
7 What do you need to collect for possible moderation? Do you need to keep a file of each pupil's written work? What records are needed of oral interactions? What records and outcomes need to be kept of practical work? Does the board provide a mark *pro-forma*? Are you expected to write an assessment of each pupil or are the mere marks sufficient?

2   The role of the teacher

Choice of entry level

A major task facing the teacher is choosing the level for which a candidate is to be entered. At first sight this does not seem to be strictly connected with the coursework, although the coursework provision and assessment may depend on it. Some assignments may be relevant and accessible to pupils of all abilities but, since differentiated tasks are central to the spirit of GCSE, it is more likely that, in order that a candidate may demonstrate success, selection of task will vary with pupils' ability. What must be borne in mind when making the choice of entry level is the whole purpose of the inclusion of coursework in GCSE.

It has at last been officially recognised that, until now, some candidates have been severely disadvantaged both by examination stress and possibly by difficulties in understanding or expressing themselves in written English. One aim of coursework is to offer such pupils the opportunity to demonstrate their abilities over a period of time and through oral communication. The second aim is to enable pupils to demonstrate skills and understandings not previously testable. You must not, therefore, allocate pupils to levels of entry based on written tests: to do so might negate the whole purpose of including coursework for assessment!

The teacher as examiner

The central problem for most teachers is how to reconcile their roles as teacher and as examiner; or, in other words, 'What help can I give?' The SEC specifically states that it is not its intention to impose extra burdens on the curriculum and so, as argued in Chapter 5 of Part 1, the submitted coursework will also form a part of the pupils' learning process. Teachers must therefore feel free to interact with pupils whenever they consider it to be necessary. Fortunately the boards recognise this fact. The degree to which it is considered important varies, however, from candidates being required to sign to say that they understand the regulations for unfair practice and that all discussions will be recorded, through recording help and adjusting marks accordingly, to keeping a 'note' of the amount of help given. MEG actually sees the help in a very positive light: 'it is expected that they will receive help and advice . . . (and that) the marks awarded will reflect the personal contribution of the candidates, including the extent to which they are able to use the advice they receive in the development of the assignments'. This latter statement has been quoted because it places the assessment procedure firmly in the learning environment. It recognises help as an integral part of learning and not as something for which pupils should be penalised. It allows the possibility that timely help can enable pupils to construct whole new areas of understanding and to achieve results which were not previously available to them.

One desirable consequence of GCSE is that all pupils will have an opportunity to experience investigative and extended ways of working. They will respond in different ways to these new challenges and for many the period of transition may be a difficult one until they have acquired confidence based on the knowledge that their strengths are being encouraged and that their ideas are valued. The teacher's role here must therefore be one of support and encouragement with no hint of the examiner showing through. The crucial advantage of continuous assessment is that there is no one occasion when you must don your examiner's hat and stand aloof from your class. What you must do, however, is be able to step outside yourself and watch the ways in which you intervene in pupils' working.

Activity 16   Monitoring your interactions with pupils

Pick any lesson with a fourth or fifth-year class and make it your aim consciously to observe and record how and when you interact with your pupils. Do you answer questions by 'telling'? Do you throw questions back to pupils with phrases such as, 'Well, what do *you* think?'? Do you vary your technique depending on whether the question is a request for a fact: 'What is the formula for solving a quadratic?'; a plea for reassurance: 'Is this right so far?'; a need for teaching: 'I don't understand why that is negative'?

You may be surprised how often you 'tell'.

Activity 17   Answering a question with a question

Take another lesson and make it your aim, just this once, to answer *every* question with a question. This is, of course, a totally false situation, but it will help sharpen your awareness of when pupils can, given encouragement, think for themselves. Pirie (1987) discusses this technique in detail.

Receiving help, as MEG recognises, can be a very constructive experience. Careful questioning can often reveal understandings and misconceptions of which you were totally unaware and which you might never have unearthed through a study of the pupils' written work. In addition, mathematical discussion with your pupils will often reveal to them errors in their own thinking. Your role as examiner demands that you do not disadvantage pupils by giving them 'told' help when what they really need is to trust their own thinking and probe a little deeper into their own understanding.

Selection of tasks

Whichever board you choose, the responsibility for selecting accessible tasks which will allow your pupils to demonstrate success can be yours, so what are the advantages and disadvantages of board-selected tasks where they are offered? Two *advantages* of the board-selected tasks are guaranteed suitability in the eyes of the board, and comparability across candidates at different centres. In addition the burden of deciding which tasks to assess is out of the hands of the teacher. *Disadvantages* lie in the area of class and task management and, in reality, suitability. If *you* are responsible for selecting the coursework to be assessed for GCSE there is no necessity for every pupil to submit the same piece of work, nor for them to do the tasks at the same time. Indeed, the work to be submitted need not necessarily be identified before it is completed; the best pieces can be selected nearer to the closing date for submission. If all pupils are to do the same board-selected assignment, however, must they all do it at the same time? Remember that coursework is *not* supposed to be a test of ability at some moment arbitrarily ordained by the board. If they do not do it together, can you give feedback to the first candidates which might unfairly advantage later ones? Remember that coursework is supposed to be part of the learning process and this entitles pupils to constructive criticism of their work. Will offering the prescribed task be seen by *all* pupils in a different light from an examination? Remember that a specific reason for putting coursework in the GCSE assessment scheme was to reduce examination-related stress.

On the question of suitability, it must be borne in mind that the boards have as little experience of setting and assessing coursework as many teachers. *You* have the advantage that you know your pupils. This would seem to indicate that you, in fact, are in a much better position to decide what assignments will enable them best to demonstrate their abilities. The boards admit that they are feeling their way in this new field, and teachers are the ideal people to offer guidance.

3 Keeping the paperwork to a minimum

There is no doubt that initially, until your board becomes comfortable with the assessment of coursework, there will be a need for a large quantity of documentation to aid moderation and validation of the awarded grades. Some careful planning in the early stages, however, can help to keep the pile of administrative paper to manageable proportions.

Activity 18   Designing a record sheet

Have in front of you the section of your board's document which lists the ways in which marks are to be awarded, and a copy of any assessment sheets on which you will have to record the final assessments.

1 Make a list of headings, if any, under which you have to report to the board, and the marks available for each heading. For example, for WJEC level 1 these would be
   - Understanding (6)
   - Strategy (3)
   - Content and development (3)
   - Communication (3)

   Even if, as with SEG, pupils do not have to demonstrate ability under all headings on each piece of work, list all the headings since you want one single *pro-forma* to cover any assignment.

2 Now list for each heading the detailed areas to be assessed and, if applicable, the marks available. This may mean that you need to reduce some of the board's rubric to a briefer reminder. For example LEAG lists criteria for the award of 2, 5 and 7 marks as follows:

Identification of the Task and Selection of a Strategy (Maximum 7 marks)

2 marks   Shows a little understanding of the task.
Strategy poorly defined.
Uses only part of the relevant information.
Uses fairly routine and/or elementary methods.
Usually explores a situation by experiment or by trial and error.

5 marks   Shows good understanding of the task.
Applies some reasoning to plan the strategy.
Adopts a systematic approach though not necessarily an efficient one.
Orders and categorises information.
Demonstrates that (s)he knows what has to be measured.
Selects appropriate variables.
Uses appropriate methods.

7 marks   Shows excellent, clear understanding of the task. Where appropriate extends the problem and/or creates sub-problems.
Applies clear reasoning to plan the strategy.
Chooses an efficient strategy.
Uses appropriate concepts and methods as the task proceeds.
Orders the information systematically and controls the variables.
Uses efficient methods to simplify the task.

This could be reduced to:

Identification of Task and Selection of a Strategy (7)
Understanding the task.
Strategy: exploration
efficient reasoning
use of relevant information
use of appropriate methods
use of appropriate concepts

3   Finally, organise your lists into a table which will fit on to one side of a sheet of A4 paper with space to fill in the marks you award. The other side can then be used for writing any comments you wish to make on an oral component or on the element of help given. Try it out. Alter it if necessary. Now duplicate this sheet so that you have one copy for each pupil for each piece of work.

4 Moderation, alternative schemes and the future

A major priority for the boards is that the school-assessed coursework should be, and be seen to be, valid, reliable and comparable across all the centres. It is reasonable to argue that addition of coursework to the examination is likely to increase the validity of the assessment since it has been introduced for the express purpose of assessing areas of achievement not easily demonstrated in a timed written paper. Reliability, too, may be increased since the coursework provides wider evidence of the pupils' achievements demonstrated on different occasions over a reasonable time-span.

Need for moderation

Comparability and, to a certain extent, validity and reliability, depend upon a workable scheme of moderation. This needs to operate at two levels: teachers within a school must agree on the standards they are using to judge the coursework, and then comparability of these standards must exist between schools.

> It is generally agreed that the teacher is uniquely placed to judge the attainment of pupils; it is also agreed that the teacher is not so well placed to judge and apply national standards of attainment. Moderation by the Board is necessary, therefore, to solve the problems of comparability of standards between teachers. Put in more concrete terms, moderation is the means by which an Examination Board's certificates are given value in the eyes of the user; it is the instrument employed by a Board in order to be able to state with confidence that standards of assessment between centres have been scrutinized and adjusted where necessary. (WJEC, GCE and GCSE Mathematics syllabus, 1986)

Forms of moderation

Unless an exceptional case arises, all boards intend to accept the rank-ordering of pupils at each level and then apply a method of moderation. Four different forms of moderation will be in use, with some boards using more than one. These fall under two main headings and are further explained below:
  (a) Moderation by inspection
  ● by the board
  ● by a consortium
  ● by consensus
  (b) Statistical moderation

(a) **Moderation by inspection**   This means that the work of a board-selected sample of candidates will be re-marked and the marks for all that centre's candidates will be adjusted on this basis. This can be conducted either by post or by a visit of the moderator to the school. It is bound to be a mammoth task for the board until general standards emerge. Recognising this, WJEC is contemplating consortium moderation where teachers are trained by the board to act as moderators in their own area under the guidance of the Chief Moderator. The SEG alternative is moderation by consensus where a group of teachers

74

meets to re-mark a sample of work and comes to an agreement on a standard set of marks.

(b) **Statistical moderation**   This is a rather different approach to the problem of comparability. It involves a statistical comparison of the coursework of a group of candidates with their work on the written timed examinations. This is spelled out in detail by NEA who will use this procedure to adjust coursework marks from a centre only if that particular centre shows a difference between its coursework average and its examination average which is markedly greater than the difference observed overall. It is not clear, however, why in such circumstances it should be the coursework mark rather than the written mark which is to be changed. Indeed this procedure may seem to negate the whole purpose of coursework assessment which is to measure different, but nonetheless valuable, abilities from the written examination. The covert message that written papers form a more accurate assessment tool than coursework can be read, too, in the NEA notes governing the examination until 1991 which state that coursework can only improve a candidate's grade from the written papers and not reduce it, and which continue with the following: *'It is expected that well-planned coursework will enhance the understanding and assimilation of the syllabus content for the written examinations.'* This is undoubtedly true, but emphatically not the reason for which coursework assessment is included in GCSE!

Activity 19   Standardising marking within your school

Using the criteria laid down by your board and the record sheet designed in Activity 18, mark two or three pieces of coursework. Ask a couple of colleagues to use the same criteria and record sheet to mark the same pieces of work. Arrange to meet when you can have at least half an hour together to compare your marks and discuss your interpretations of the criteria. You may feel you need to do this exercise several times before you are confident that the whole department has the same approaches and standards.

Alternative schemes

This book so far has been concerned with the main 'mode 1' syllabus available from the five GCSE boards. At the time of writing, however, there are several other schemes in the process of seeking approval by SEC and these are discussed, in broad outline, here. It is possible to offer an alternative syllabus as either a 'mode 1' (external examination based on a board's published syllabus), 'mode 2' (external examination based on a syllabus submitted by a centre or group of centres) or 'mode 3' (an internally assessed scheme based on a syllabus submitted by a centre or group of centres) and all three options are under preparation.

SMP

SMP (School Mathematics Project) 11–16 has a 'mode 1' syllabus available from MEG in association with LEAG for 1988 based on their course of booklets and colour-coded books. One of the advantages of this scheme for teachers in schools already using the course is that they will have little difficulty selecting which level of examination entry is appropriate for each pupil. The structure of the curriculum materials makes this choice relatively obvious. The coursework tasks to be undertaken and the guidelines for assessment are also very prescriptive, offering strong support for the less confident teacher. Twenty-five per cent of the total marks are for coursework and 5% for oral tests. The syllabus, however, does not cover all the content prescribed in the National Criteria, hence its experimental status.

GAIM

GAIM (Graded Assessment in Mathematics) is also preparing a 'mode 1' submission to SEC, initially for pilot schools only, but to be offered nationally from 1991 if acceptable to SEC. The major feature of the scheme is that assessment is based on 100% coursework. It is available from LEAG and provides a detailed cumulative record of what each student achieves in mathematics. This record has two elements:

(a) **open-ended coursework assignments**, including practical problems, investigations and extended pieces of work,

(b) **topic criteria** at each of 15 levels of attainment which are continuously assessed as part of normal coursework.

It can be used to complement any mathematics scheme which is in the spirit of the Cockcroft Report. Its arrangement of graded levels of achievement, through which a pupil progresses, means that the question of examination entry levels does not arise and the choice of coursework is a matter for both the pupil and teacher as the course proceeds, although specimen investigations, practical problem-solving tasks and starting points for extended pieces of work are available from GAIM together with associated assessment schedules. There is no percentage weighting for the various components. To be awarded a particular grade the pupil must have achieved the corresponding level of attainment in *each* of the four components:

- practical problem solving
- investigations
- extended pieces of work
- topic criteria

All 'mode 1' proposals are very thoroughly scrutinised by SEC and its attitude to these two proposals is that although neither fulfils all the National Criteria, either from the curriculum development or from the assessment point of view, each has interesting aspects and for this reason has been given a 'limited life' within which to develop.

MEI

MEI (Mathematics in Education and Industry) has a 'mode 2' examination offered jointly by MEG and SEG from 1988. The unusual feature here is that there are four levels (not three) of entry:

- Foundation   – available grades: E–G
- Intermediate – available grades: C–F
- Standard   – available grades: B–E
- Higher   – available grades: A–D

The scheme has evolved from the long established MEI O-level and from a 'mode 3' CSE developed by a group of schools in the south-west of England. Nineteen per cent of the total marks are for the coursework including an oral test to measure 'the candidate's ability to discuss mathematics', and a further 7% for mental calculations. Tasks are available from MEI but schools are encouraged to move on to writing their own material.

ATM

A number of schools and groups of centres have approached SEC with mode 3 proposals but many of these have been turned down because they do not fit the National Criteria. ATM is working on a radically different approach to assessment at GCSE. They have devised their own aims and assessment objectives which, it is claimed, subsume the National Criteria. There is a noticeable emphasis on the ability of pupils to work cooperatively, develop autonomy and evaluate their own work. Like GAIM, the scheme is assessed 100% by coursework, with candidates submitting a 'folio of work gathered over the period of the course'. This is assessed by the pupil's teacher, and ATM is overtly aware of the danger of setting coursework which does not fit the ability of the candidate. 'Care must be taken that all students are set tasks to which they can make an initial response, but which are challenging enough to extend them; thus all students are stretched and none daunted'. This problem is highlighted in some of the tasks in Part 2. The interesting feature of the scheme is the intention that schools will work through ATM, using their syllabus as a framework, to research and design their own course and assessment scheme. The necessary 'enthusiastic and sympathetic examination group' who will support this venture is likely to be SEG.

Modular courses

A learning environment not yet considered in this book is that of the schools who use a modular approach to teaching. There is an increasing number of these, and the standard 'mode 1' schemes may not be a suitable way in which to assess such pupils. The major boards are looking at this problem, but until now there is no agreement on the length of a module or on the number to be assessed to allow certification. MEG talks of a 'specified number of modules which together offers sufficient depth, breadth, coherence and quality to make the programme suitable for single subject certification'. It seems possible, however, to design hybrid modules such as 'electronics and music' or 'mathematics and home economics'. What a modular scheme must provide is modules which are assessed while they are being studied, and a programme consisting of a number of modules which is accredited at its conclusion. The problem of the maturation of pupils, mentioned earlier, is particularly acute here.

Although the majority of candidates for GCSE will be in schools, there will be a sizeable number entering from Colleges of Further Education. These may be mature students wanting to get further qualifications or people wanting to re-sit examinations to obtain better grades. Both of these categories will expect to spend only one year preparing for their assessment. For such people various bodies are trying to devise specialised modular courses. One noteworthy feature of all the schemes so far submitted is that they do not award a grade G. The most common approach is to provide a package of modules split into separate groups so that if candidates select the prescribed number of modules from the various groups they will have covered the National Criteria for mathematics and can be awarded a GCSE grade. These modules will encompass areas such as 'Maths for banking', 'Maths for surveying', 'Maths for catering' in addition to more pure mathematical fields. The ideas are really, however, still in their infancy.

And what of the future?

One current concern of SEC is the lack of guidance given by the boards on their coursework components. For some teachers this may be a valuable freedom, enabling them to shape their assessment in a way which will reflect the abilities of their pupils. For others this may cause anxiety that they do not know how to present their pupils' strengths to the board. The boards' answer to this dilemma is to initiate board-provided in-service training. This still leaves concern as to how often and for how long this will be available. What about schools which do not opt to do the coursework component until 1991? How will schools who have misinterpreted a board's scheme be dealt with? Can the moderation methods combat erratic assessment? These are all problems which the boards will have to consider.

In Chapter 4 of Part 1 the idea of grade-related criteria was discussed. Examination boards have been asked to assess how far current syllabuses and examination questions fit with the SEC draft grade criteria. The answer seems to be that the boards are in favour of grade criteria but in their present form they are unworkable. One way ahead for SEC would be to move away from the idea of a set of general grade criteria towards more explicit topic-specific criteria.

One fact is abundantly clear: over the next few years the assessment of all subjects, but most especially mathematics, will evolve from the present experimental beginnings, and the teacher is in a powerful position to influence that evolution. If you wish to make an input into the question of how pupils should be assessed at 16-plus, now is the moment to speak.

Good luck!

Appendices

Appendix 1 LEAG guidelines

Only criteria for the award of 7 marks, 5 marks and 2 marks are given for the first two categories and criteria for the award of 6 marks, 4 marks and 1 mark for the third category. Teachers should use their professional judgement in awarding intermediate marks.

Identification of the task and selection of a strategy (maximum 7 marks)

2 marks
Shows a little understanding of the task.
Strategy poorly defined.
Uses only part of the relevant information.
Uses fairly routine and/or elementary methods.
Usually explores a situation by experiment or by trial and error.

5 marks
Shows good understanding of the task.
Applies some reasoning to plan the strategy.
Adopts a systematic approach though not necessarily an efficient one.
Orders and categorises information.
Demonstrates that (s)he knows what has to be measured.
Selects appropriate variables.
Uses appropriate methods.

7 marks
Shows excellent, clear understanding of the task. Where appropriate extends the problem and/or creates sub-problems.
Applies clear reasoning to plan the strategy.
Chooses an efficient strategy.
Uses appropriate concepts and methods and develops the methods as the task proceeds.
Orders the information systematically and controls the variables.
Uses efficient methods to simplify the task.

Implementation and reasoning (maximum 7 marks)

2 marks
Processes some data. Some simple calculations completed accurately.
Uses the information provided.
Recognises some simple patterns.
Continues and extends these simple patterns.
Produces little or nothing in the way of general rules.
Produces some sketches and graphs where applicable.

5 marks
Processes data accurately and makes only minor errors.
Applies some variety of skills, knowledge and procedures to a problem.
Makes conjectures about patterns, etc, and tests them.
Attempts to formulate some general rules.
Devises simple formulae where appropriate.
Recognises patterns.
Makes some use of symbols when generalising.
Attempts to verify and justify results.
Makes use of a range of resources.

7 marks   Processes data very accurately.
Discriminates between necessary and redundant information.
Plans and schedules a range of relevant mathematical tasks.
Applies a variety of skills, knowledge and procedures to a problem.
Makes and tests conjectures.
Formulates general rules.
Recognises patterns.
Where appropriate, makes use of symbols when generalising.
Devises, tests and uses a simple mathematical model.

Interpretation and communication (maximum 6 marks)

1 mark   Produces a limited attempt to relate results to the original problem.
Summarises the results and makes some valid observations.
Describes some patterns or features of the results.

4 marks   States results achieved and relates these to the original problem but usually without being able to state many valid conclusions.
Gives a clear account of the task but without giving reasons for the strategies used and/or explaining the assumptions made.
Presents results in an orderly sequence.
Uses an adequate range of mathematical language and symbols, including appropriate visual forms.

6 marks   States results achieved and draws and states valid conclusions.
Gives a clear account of the task, giving reasons for the strategies used and explaining assumptions made.
Selects the most appropriate methods for communicating results.
Makes effective use of a range of mathematical language and notation, diagrams, charts and, where appropriate, computer output.

Appendix 2 MEG guidelines

Classification of assessment	Maximum mark	Guidance for marking
Overall design and strategy	4	4 – A well-defined problem, appropriate use of techniques, well-stated conclusions, strong personal contribution
		2 – Routine approach, satisfactory techniques, some statement(s) of conclusion(s), average help needed
		0 – Trivial or poorly stated problem, unsuitable techniques, a lack of conclusion, even with considerable help
Mathematical content	4	4 – Commendable use of concepts and methods showing a good range of knowledge; development of these concepts and methods as the work progresses
		2 – Appropriate concepts and methods without development or refinement, showing competence in a limited range of techniques
		0 – Inadequate for the assignment
Accuracy	4	4 – Careful and accurate work including, where appropriate, computation, manipulation, construction and measurement with correct units
		2 – Some errors, but not sufficient to invalidate the work
		0 – Inaccurate work
Clarity of argument and presentation	4	4 – A clearly-expressed contribution with effective use of mathematical language, symbols, conventions, tables, diagrams, graphs etc
		2 – Adequate presentation, average use of appropriate language, symbols, conventions etc
		0 – Disorganised, untidy work, poorly expressed
Controlled element	4	4 – Demonstrates knowledge and understanding of the assignment, can apply the learning in a different situation.
		2 – Has shown some learning of mathematics, with a limited ability to use it.
		0 – The assignment has not contributed to mathematical learning.

Appendix 3 NEA guidelines

The candidate's work should be assessed at frequent and appropriate intervals during the course. The maintenance of a record of these periodic assessments, irrespective of the type of work being undertaken, will enable the teacher to build up a picture of the candidate's ability to assist in the final assessment. The final assessment is to be expressed in the range 0–100. In each assessment set a candidate's marks should reflect the level of attainment demonstrated.

The following assessment guide outlines the main points which the teacher should consider and the mark ranges which can be used when making the final assessment of the candidate's work.

(a) **Practical work (total 30)**
    (i) Planning, execution and completion of the work and presenting accurate results. (0–12)
    (ii) Knowledge and understanding of the processes associated with the use of equipment. (0–12)
    (iii) Skill in using instruments and equipment. (0–6)

(b) **Investigational work (total 40)**
    (i) Planning and preparation of the work. (0–7)
    (ii) Variety and sources of information. (0–5)
    (iii) Clarity and conciseness of expression. (0–5)
    (iv) Originality of approach and presentation. (0–5)
    (v) Ability to draw valid conclusion. (0–10)
    (vi) Ability to carry out work of an extended nature. (0–8)

(c) **Assimilation (total 30)**
    (i) Recall of knowledge. (0–3)
    (ii) Oral communication concerning the subject matter. (0–7)
    (iii) Understanding of the subject matter and concepts involved. (0–10)
    (iv) Awareness of relevance and application of the subject matter. (0–10)

Appendix 4 SEG guidelines

Guidelines for coursework assessment

Teachers should consider the following questions when assessing their candidates' work. Not all sections, or questions within sections, will be appropriate to every piece of work, but the Centre based assessment in its totality should cover most of them.

The sections and questions given are not expected to be encountered in any particular order in a candidate's work.

1 Comprehension of task
Did the candidate understand what the task involved?
Has the candidate asked appropriate questions?
Can the candidate form suitable problems?

2 Planning
(a) Strategy
Did the candidate plan an overall strategy?
(b) Choice of method
Did the candidate consider a range of methods?
Did the candidate select a suitable method?
(c) Choice of information required
Did the candidate consider which information he/she might need to start the problem?
(d) Equipment
Did the candidate select appropriate equipment/material?

3 Carrying out the task
(a) Accuracy
Was the candidate accurate in his/her work?
Did he/she work to an appropriate degree of accuracy?
Where appropriate, were checks made?
(b) Amendments to strategy
Was the candidate aware of the implications of results at each stage?
Did the candidate make appropriate decisions in the light of results?
(c) Patterns
Did the candidate look for patterns?
Did the candidate recognise patterns where they existed?
(d) Conjecture
Was the candidate able to predict possible outcomes for
(i) the particular problems?
(ii) other, related problems?
Did the candidate test conjectures?
(e) Generalisation
Did the candidate attempt to generalise?
Did he/she find any generalisations?
Did the candidate express generalisations correctly in
(i) words?
(ii) mathematical symbols?
Did the candidate use appropriate checks to test the generalisations found?
(f) Depth of study
Did the candidate carry through the task to an appropriate point?

4 Communication

(a) Clarity

Did the candidate explain clearly his/her actions at each stage?

Did the candidate use appropriate methods to record data?

Did the candidate use appropriate methods to present results?

(b) Justification

Did the candidate justify his/her results?

(c) Evaluations

Did the candidate attempt to evaluate results?

Were those evaluations appropriate?

Was the candidate aware of possible future developments? If so, were appropriate suggestions made?

Guidelines for oral assessment

1 The following notes are offered as a guide to help teachers structure their oral assessment. The questions shown within each section are descriptive and not prescriptive. The actual questions asked should be directly relevant to the work being assessed. Sometimes it may not be relevant to ask questions from all sections. Questions should be asked at appropriate stages during the candidate's work.

2 Questions should generally be open ended so as to allow the candidate the greatest opportunity to show what has been learnt. Closed questions, however, may be required to obtain clarification of answers given. Teachers are warned of the danger of asking leading questions, i.e. questions likely to bias an answer. The following are offered as examples of suitable questions:

(a) Starting points and assumptions

(i) Were there any things you had to find out before the work could be started?

(ii) What things did you assume? (Rooms being rectangular; lines being parallel, etc.)

(iii) What sort of accuracy did you think would be needed? How did you decide this?

(iv) Were there any checks you had to make? (Zeroing scales before use, making measurements twice, questionnaire not biased, etc.)

(b) Method of recording and reporting

(i) How did you record your information?

(ii) Why did you choose this method?

(iii) Did you consider any other methods?

(iv) What information did you want your results to convey?

(v) Did you think of any other ways of doing this?

(vi) Did you consider . . . method?

(c) Results

(i) How did you reach your results?

(ii) What do they mean?

(iii) Were they the ones you expected? Why not?

(iv) Are there any other things you could say from your work which are not in the report?

(d) Checking

(i) Tell me how you checked your work.

(ii) Were there any things you had to correct?

(iii) Were there any things you now feel you ought to have checked?

(iv) Did you check . . .?

(e) False leads

(i) Were there any parts of your work which you started and then had to scrap and start again? Why?

(ii) Did this (lead) help with any other parts of the work?

(iii) Were there any ways you could have continued along the same path?

(f) Extension

(i) Can you think of any ways you could extend the work you have done into other areas of mathematics?

      (ii) Which areas of mathematics would you need to use?

      (iii) Are there any things you cannot do at the moment which you would need to be able to do to follow up your ideas?

      (iv) Suppose (something) had been (something else), what effect would this have had on your work?

(g) Conclusions

      (i) Can you describe what you have found out doing your work?

      (ii) Are there any things which you did not know how to do before you started which you can do now?

(1988 Examination)

Appendix 5 WJEC marking scheme

Levels 2 and 3

The practical investigation:

The Assignment:
Understanding	3
Strategy	5
Content and development	6
Communication	7

Continuous assessment:
Understanding	3
Method	3
Conclusion	3

Aural test	10
Total	40

The problem solving investigation

The Assignment:
Understanding	3
Strategy	5
Content and development	6
Communication	7

Continuous assessment:
Understanding	3
Method	3
Conclusion	3
Total	30

Level 1

The investigation:

The assignment:
Understanding	6
Strategy	3
Content and development	3
Communication	3

Continuous assessment:	15
Total	30

The practical task
A series of **three** practical exercises (10 marks for each exercise)	30
Aural test	10
Total	40

Criteria for awarding marks for internal assesment

Levels 2 and 3

Assessment categories

Understanding

Shows evidence of understanding the nature of the problem.	1
Understands the nature of the problem, identifies important features or variables.	2
Understands the nature of the problem, identifies important features or variables, extends and creates related problems.	3

Strategy

Identifies the areas for investigation and attempts to formulate the problem in terms of appropriate questions to be answered.   1

Identifies the areas for investigation and attempts to formulate the problem in terms of appropriate questions to be answered. Attempts to collect data.   2

Identifies the areas for investigation and formulates the problem in terms of appropriate questions to be answered. Collects data which will enable answers to be formulated.   3

Clearly identifies the areas for investigation and formulates the problem in terms of appropriate questions to be answered. Collects data which will enable answers to be formulated. Attempts to identify additional lines of enquiry and form techniques for comparing.   4

Clearly identifies the areas for investigation and formulates the problem in terms of appropriate questions to be answered. Collects data which will enable answers to be formulated. Identifies additional lines of inquiry and forms techniques for comparing.   5

Content and development

(a) Shows some ability in using appropriate methods to represent the data collected and in making relevant calculations leading to the drawing of conclusions.   1

Shows an ability to choose appropriate methods to represent the data collected and to make relevant calculations leading to the drawing of conclusions based on sound deduction.   2

Shows an ability to choose appropriate methods to represent the data collected and to make relevant calculations leading to the drawing of conclusions based on sound deduction. Attempts to make appropriate use of mathematics in identifying alternative lines of inquiry and in the interpretation of results.   3

Shows an ability to choose appropriate methods to represent the data collected and to make relevant calculations leading to the drawing of conclusions based on sound deduction. Appropriate use of mathematics in identifying additional lines of inquiry and in the interpretation of results. Reasons logically and precisely.   4

(b) Generally accurate and appropriate work, supported by computation, reference and appropriate diagrammatic work.   1

Accurate, appropriate and concise work well supported by computation, reference and appropriate diagrammatic work.   2

Communication

(a) Produces some conclusions with a suitable selection of material.   1

Produces some conclusions with suitable selection of material and some suggestions for extensions to the investigation.   2

Produces clearly stated conclusions with a suitable selection of material and suggestions for extensions and modifications to the investigation.   3

Produces clearly stated conclusions with a suitable selection of material and suggestions for extensions or modifications to the investigation. Recognises and states clearly to what extent the results are valid or incomplete and discusses relevant areas for further investigation.   4

(b) Clearly and accurately expressed work.   1

Clearly and accurately expressed work with some use of mathematical language and concepts to present conclusions.   2

Clearly and accurately expressed work, making a wide and effective use of mathematical language and concepts to present conclusions logically and concisely.   3

Continuous assessment

This assessment plan covers the candidate's ability to explain the methods and strategies orally by responding to questions both during the investigation period and on presentation of the final work.

Assessment area	*Possible questions*	*Marking guidelines*	
Understanding of the problem	What do you think the problem is about? What will you have to take into account?	Shows some understanding of the problem.	1
		Has a good grasp of the problem and is able to explain some of the constraints	2
		Shows clear understanding of the problem and a full appreciation of the major constraints.	3
Method	What have you done so far? What will you do next? Do you have any difficulties? Can you explain what they are?	Has some appreciation of the difficulties.	1
		Proceeds without difficulty and gives an outline of the work undertaken and the intended next steps.	2
		Is able to explain the nature of any problem area and is able to ask for appropriate help.	3
Conclusion (after the teacher has read the finished work)	What did your initial investigation show? What did you find out? Did you investigate any other areas? What did you find out?	Has obtained some results.	1
		Has obtained some results and can explain some conclusions.	2
		Can express the conclusions clearly and explain any remaining difficulties, making effective use of appropriate mathematical language.	3

Level 1

Assessment categories

Understanding

(a) Shows evidence of understanding the nature of the problem.   1

Understands the nature of the problem and identifies some of the important features.   2

Understands the nature of the problem and identifies important features.   3

(b) Is able to make progress but with considerable teacher guidance.   1

Has ability to make progress with some teacher guidance.   2

Strong personal contribution, requires little teacher assistance.   3

Strategy

Identifies the areas for investigation and attempts to formulate the problem in terms of appropriate questions to be answered. Attempts to collect data.   1

Identifies the areas for investigation and attempts to formulate the problem in terms of appropriate questions to be answered.   2

Identifies the areas for investigation and formulates the problem in terms of appropriate questions to be answered. Collects data and interprets it.   3

Content and development

Shows some ability in using appropriate methods to represent data collected.   1

Shows some ability to choose appropriate methods to represent data collected and
make relevant calculations. With suitable guidance, is able to make appropriate use
of mathematics in the interpretation of results.   2

Shows an ability to choose appropriate methods to represent data collected and make
relevant calculations. Attempt to make appropriate use of mathematics in the
interpretation of results.   3

Communication

Clearly expressed work.   1

Clearly expressed work with some use of mathematical language to present conclusions.   2

Clearly expressed work making use of mathematical language to present conclusions.   3

Appendix 6 The National Criteria for Mathematics

1 Introduction

1.1 This statement of National Criteria for Mathematics sets out the essential requirements which must be satisfied by all syllabuses for examinations entitled Mathematics.

It will be the responsibility of Examining Groups to determine their own syllabuses and techniques of assessment in accordance with these criteria.

1.2 When devising examinations in Mathematics regard should be paid to the discussion and recommendations to be found in the Report of the Committee of Inquiry into the Teaching of Mathematics in Schools (the Cockcroft Report). Of especial importance is the statement that "pupils must not be required to prepare for examinations which are not suited to their attainment nor must these examinations be of a kind which will undermine the confidence of pupils".

Mathematics examinations at 16+ exert considerable influence on the content and pace of work in secondary classrooms. It is therefore essential that syllabuses and methods of assessment for these examinations should not conflict with the provision and development of appropriate and worthwhile mathematics courses in schools.

1.3 These criteria should be so interpreted that any scheme of assessment will:

(i) assess not only the performance of skills and techniques but also pupils' understanding of mathematical processes, their ability to make use of these processes in the solution of problems and their ability to reason mathematically;

(ii) offer differentiated examination papers so that, by choosing papers at an appropriate level, pupils are enabled to demonstrate what they know and can do rather than what they do not know and cannot do;

(iii) encourage and support the provision of courses which enable pupils to develop their knowledge and understanding of mathematics to the full extent of their capabilities, to have experience of mathematics as a means of solving practical problems and to develop confidence in their use of mathematics;

(iv) have regard to the need for examination tasks to relate, where appropriate, to the use of mathematics in everyday situations.

2 Aims

The statement which follows sets out ideal educational aims for all those following courses in Mathematics which lead to GCSE examinations. Some of these aims refer to the development of attributes and qualities which it might not be possible, or desirable, to assess directly.

All courses should enable pupils to:

2.1 develop their mathematical knowledge and oral, written and practical skills in a manner which encourages confidence;

2.2 read mathematics, and write and talk about the subject in a variety of ways;

2.3   develop a feel for number, carry out calculations and understand the significance of the results obtained;

2.4   apply mathematics in everyday situations and develop an understanding of the part which mathematics plays in the world around them;

2.5   solve problems, present the solutions clearly, check and interpret the results;

2.6   develop an understanding of mathematical principles;

2.7   recognise when and how a situation may be represented mathematically, identify and interpret relevant factors and, where necessary, select an appropriate mathematical method to solve the problem;

2.8   use mathematics as a means of communication with emphasis on the use of clear expression;

2.9   develop an ability to apply mathematics in other subjects, particularly science and technology;

2.10   develop the abilities to reason logically, to classify, to generalise and to prove;

2.11   appreciate patterns and relationships in mathematics;

2.12   produce and appreciate imaginative and creative work arising from mathematical ideas;

2.13   develop their mathematical abilities by considering problems and conducting individual and cooperative enquiry and experiment, including extended pieces of work of a practical and investigative kind;

2.14   appreciate the interdependence of different branches of mathematics;

2.15   acquire a foundation appropriate to their further study of mathematics and of other disciplines.

3 Assessment objectives

The objectives which follow set out essential mathematical processes in which candidates' attainment will be assessed. They form a minimum list of qualities, abilities and skills. The weight attached to each of these objectives may vary for different levels of assessment within a differentiated system.

Any scheme of assessment will test the ability of candidates to:

3.1   recall, apply and interpret mathematical knowledge in the context of everyday situations;

3.2   set out mathematical work, including the solution of problems, in a logical and clear form using appropriate symbols and terminology;

3.3   organise, interpret and present information accurately in written, tabular, graphical and diagrammatic forms;

3.4   perform calculations by suitable methods;

3.5   use an electronic calculator;

3.6   understand systems of measurement in everyday use and make use of them in the solution of problems;

3.7   estimate, approximate and work to degrees of accuracy appropriate to the context;

3.8   use mathematical and other instruments to measure and to draw to an acceptable degree of accuracy;

3.9   recognise patterns and structures in a variety of situations, and form generalisations;

3.10   interpret, transform and make appropriate use of mathematical statements expressed in words or symbols;

3.11   recognise and use spatial relationships in two and three dimensions, particularly in solving problems;

3.12   analyse a problem, select a suitable strategy and apply an appropriate technique to obtain its solution;

3.13   apply combinations of mathematical skills and techniques in problem solving;

3.14   make logical deductions from given mathematical data;

3.15   respond to a problem relating to a relatively unstructured situation by translating it into an appropriately structured form.

Two further assessment objectives can be fully realised only by assessing work carried out by candidates in addition to time-limited written examinations. From 1988 to 1990 all Examining Groups must provide

at least one scheme which includes
some elements of these two objectives.
From 1991 these objectives must be
realised fully in all schemes.

3.16   respond orally to questions about
mathematics, discuss mathematical
ideas and carry out mental
calculations;

3.17   carry out practical and investigational
work, and undertake extended
pieces of work.

Appendix 7 SEG assessment objectives

The examination will assess a candidate's ability to:

(a) Knowledge
   (i) recall mathematical knowledge in the context of everyday situations;
   (ii) understand systems of measurement in everyday use.

(b) Skills
   (i) set out mathematical work, including the solution of problems, in a logical and clear form using appropriate symbols and terminology;
   (ii) interpret mathematical knowledge in the context of everyday situations;
   (iii) organise, interpret and present information accurately in written, tabular, graphical and diagrammatic forms;
   (iv) perform calculations by suitable methods;
   (v) use an electronic calculator;
   (vi) estimate, approximate and work to degrees of accuracy appropriate to the context;
   (vii) use mathematical and other instruments to measure and to draw to an acceptable degree of accuracy.

(c) Applications
   (i) apply mathematical knowledge and skills in the context of everyday situations;
   (ii) make use of systems of measurement in the solution of problems;
   (iii) interpret, transform and make appropriate use of mathematical statements expressed in words or symbols.

(d) Problem solving
   (i) recognise patterns and structures in a variety of situations, and form generalisations;
   (ii) recognise and use spatial relationships in two and three dimensions, particularly in solving problems;
   (iii) analyse a problem, select a suitable strategy and apply an appropriate technique to obtain its solution;
   (iv) apply combinations of mathematical skills and techniques in problem solving;
   (v) make logical deductions from given mathematical data;
   (vi) respond to a problem relating to a relatively unstructured situation by translating it into an appropriately structured form.

(1988 Examination)

Appendix 8 NISEC assessment scheme

% Marks – *Course:*	20
Exam:	70
Other:	Award for computation test 10

Papers

Three different pairs of papers
Basic: Grades E–G
Intermediate: Grades C–F
High: Grades A–D

Coursework assignments

Four assignments (one extended piece of work may count as two).
At least one from each of the two areas:

1 Practical geometry/Measurement/Everyday applications/ Statistical work.

2 Pure mathematics investigations.

Board prescription and guidance on task choice

'Teachers are responsible for devising the assignments'. The total assessed work should cover opportunities to:

(a) organise and interpret information;
(b) collect and select data/measurements;
(c) select and carry out appropriate calculations;
(d) organise and solve problems;
(e) check, interpret and evaluate results;
(f) recognise patterns, propose generalisations (and explore their validity);
(g) explain methods/strategies adopted.

'It is essential that candidates be set tasks which will enable them to demonstrate what they know, understand and can do.' Teachers should select only assignments which are within the capabilities of the particular candidate. Group assignments are permitted if individual contributions can be reliably assessed.

Allocation of marks

	Basic	*Int.*	*High*
Comprehension of task and planning:	3	5	7
Examination of task:	3	5	7
Communication and Evaluation:	3	5	7
Maximum possible: (each scaled to 20%)	36	60	84

Board prescription and guidance on marking	Candidates are expected to do no more than four assignments altogether and teachers will select for formal assessment those likely to obtain the highest marks. General guidelines are given in the form of descriptions of the three categories for marks, but 'teachers will be required to attend training workshops organised by the Council at which they will be given detailed information regarding the assessment of assignments'.
Moderation	Moderation by the Council's Coursework Moderator.

Bibliography

Bell A., Rooke D. and Wigley A. (1978), *Journey into Maths*, South Notts
    Project, Blackie, London
Cockcroft W. H. (1982), Chairman of the Committee of Inquiry into the
    Teaching of Mathematics in Schools, *Mathematics Counts*, HMSO, London
Forster J. and Wardle M. (1986), *Countdown to GCSE: Mathematics*, Macmillan
    Education, Basingstoke
JCNC (1985), *National Criteria for Mathematics*, GCE and CSE Boards' Joint
    Council for 16+ National Criteria
JMB (1985), *The Language of Functions and Graphs*, Joint Matriculation
    Board/Shell Centre for Mathematical Education, Manchester
Pirie S. E. B. (1987), *Mathematical Investigations in Your Classroom*, Macmillan
    Education, Basingstoke
SEC (1985), *Working Paper 2. Coursework Assessment in GCSE*, Secondary
    Examinations Council, London
SEC (1986), *Mathematics. GCSE. A Guide for Teachers*, Open University Press,
    Milton Keynes